Wittgenstein on Mathematics

This book offers a detailed account and discussion of Ludwig Wittgenstein's philosophy of mathematics. In Part I, the stage is set with a brief presentation of Frege's logicist attempt to provide arithmetic with a foundation and Wittgenstein's criticisms of it, followed by sketches of Wittgenstein's early views of mathematics, in the *Tractatus* and in the early 1930s. Then (in Part II), Wittgenstein's mature philosophy of mathematics (1937–44) is carefully presented and examined. Schroeder explains that it is based on two key ideas: the *calculus view* and the *grammar view*. On the one hand, mathematics is seen as a human activity—calculation—rather than a theory. On the other hand, the results of mathematical calculations serve as grammatical norms. The following chapters (on mathematics as grammar; rule-following; conventionalism; the empirical basis of mathematics; the role of proof) explore the tension between those two key ideas and suggest a way in which it can be resolved. Finally, there are chapters analysing and defending Wittgenstein's provocative views on Hilbert's formalism and the quest for consistency proofs and on Gödel's incompleteness theorems.

Severin Schroeder is Associate Professor of Philosophy at the University of Reading. He has published three monographs on Wittgenstein: *Wittgenstein: The Way Out of the Fly Bottle* (2006), *Wittgenstein Lesen* (2009), and *Das Privatsprachen-Argument* (1998). He is the editor of *Wittgenstein and Contemporary Philosophy of Mind* (2001) and *Philosophy of Literature* (2010).

Wittgenstein's Thought and Legacy
Edited by Eugen Fischer
University of East Anglia, UK
Severin Schroeder
University of Reading, UK

Wittgenstein on Thought and Will
Roger Teichmann

Wittgenstein on Sensation and Perception
Michael Hymers

Wittgenstein and Naturalism
Edited by Kevin M. Cahill and Thomas Raleigh

Wittgenstein on Mathematics
Severin Schroeder

For more information about this series, please visit: www.routledge.com/Wittgensteins-Thought-and-Legacy/book-series/WTL

Wittgenstein on Mathematics

Severin Schroeder

Routledge
Taylor & Francis Group
NEW YORK AND LONDON

First published 2021
by Routledge
52 Vanderbilt Avenue, New York, NY 10017

and by Routledge
2 Park Square, Milton Park, Abingdon, Oxon, OX14 4RN

Routledge is an imprint of the Taylor & Francis Group, an informa business

Library of Congress Cataloging-in-Publication Data
Names: Schroeder, Severin, author.
Title: Wittgenstein on mathematics / Severin Schroeder.
Description: New York, NY : Routledge, 2021. | Series: Wittgenstein's thought and legacy | Includes bibliographical references and index.
Identifiers: LCCN 2020037371 (print) | LCCN 2020037372 (ebook) | ISBN 9781844658626 (hardback) | ISBN 9781003056904 (ebook)
Subjects: LCSH: Mathematics—Philosophy.
Classification: LCC QA8.4 .S376 2021 (print) | LCC QA8.4 (ebook) | DDC 510.1—dc23
LC record available at https://lccn.loc.gov/2020037371
LC ebook record available at https://lccn.loc.gov/2020037372

ISBN: 978-1-844-65862-6 (hbk)
ISBN: 978-1-003-05690-4 (ebk)

Typeset in Sabon
by Apex CoVantage, LLC

Contents

Preface viii
List of Abbreviations xii

PART I
Background 1

1 **Foundations of Mathematics** 3

2 **Logicism** 9
2.1 *Frege's Logicism* 9
2.2 *The Class Paradox and Russell's Theory of Types* 12
2.3 Tractatus Logico-Philosophicus*: Logicism Without Classes* 13

3 **Wittgenstein's Critique of Logicism** 15
3.1 *Can Equality of Number Be Defined in Terms of One-to-One Correlation?* 15
3.2 *Frege's (and Russell's) Definition of Numbers as Equivalence Classes Is Not Constructive: It Doesn't Provide a Method of Identifying Numbers* 21
3.3 *Platonism* 22
3.4 *Russell's Reconstructions of False Equations Are Not Contradictions* 26
3.5 *Frege's and Russell's Formalisation of Sums as Logical Truths Cannot Be Foundational as It Presupposes Arithmetic* 27
3.6 *Even If We Assumed (for Argument's Sake) That All Arithmetic Could Be Reproduced in Russell's Logical Calculus, That Would Not Make the Latter a Foundation of Arithmetic* 31

4 **The Development of Wittgenstein's Philosophy of Mathematics: *Tractatus* to *The Big Typescript*** 35
4.1 Tractatus Logico-Philosophicus *35*
4.2 Philosophical Remarks *(MSS 105–8: 1929–30) to* The Big Typescript *(TS 213: 1933) 36*

PART II
Wittgenstein's Mature Philosophy of Mathematics (1937–44) 55

5 **The Two Strands in Wittgenstein's Later Philosophy of Mathematics** 57

6 **Mathematics as Grammar** 59

7 **Rule-Following** 78
7.1 Rule-Following and Community 88

8 **Conventionalism** 93
8.1 Quine's Circularity Objection 95
8.2 Dummett's Objection That Conventionalism Cannot Explain Logical Inferences 101
8.3 Crispin Wright's Infinite Regress Objection 103
8.4 The Objection to 'Moderate Conventionalism' From Scepticism About Rule-Following 105
8.5 The Objection From the Impossibility of a Radically Different Logic or Mathematics 109
8.6 Conclusion 124

9 **Empirical Propositions Hardened Into Rules** 126
Synthetic A Priori 134

10 **Mathematical Proof** 141
10.1 What Is a Mathematical Proof? 142
(a) Proof That $a^0 = 1$ 150
(b) Skolem's Inductive Proof of the Associative Law of Addition 150
(c) Cantor's Diagonal Proof 151
(d) Euclid's Construction of a Regular Pentagon 158

(e) Euclid's Proof That There Is No Greatest Prime Number 160
(f) Proof (Calculation) in Elementary Arithmetic 166
Proof and experiment 169
10.2 What Is the Relation Between a Mathematical Proposition and Its Proof? 171
10.3 What Is the Relation Between a Mathematical Proposition's Proof and Its Application? 181

11 Inconsistency 189

12 Wittgenstein's Remarks on Gödel's First Incompleteness Theorem 203
12.1 Wittgenstein Discusses Gödel's Informal Sketch of His Proof 206
12.2 'A Proposition That Says About Itself That It Is Not Provable in P' 207
12.3 The Difference Between the Gödel Sentence and the Liar Paradox 209
12.4 Truth and Provability 210
12.5 Gödel's Kind of Proof 213
12.6 Wittgenstein's First Objection: A Useless Paradox 216
12.7 Wittgenstein's Second Objection: A Proof Based on Indeterminate Meaning 218

13 Concluding Remarks: Wittgenstein and Platonism 220

Bibliography 226
Index 234

Preface

This book aims to offer a coherent account of Wittgenstein's mature philosophy of mathematics. As his later philosophy of language and mind presented in *Philosophical Investigations* is vastly superior to the ideas developed in his first book *Tractatus Logico-Philosophicus*—indeed his later work can be seen as the result of a profound understanding of the shortcomings of his early work—similar cognitive progress might be expected, and can I think be found, in Wittgenstein's thinking about mathematics. It would certainly appear that Wittgenstein himself regarded the extensive changes and developments of his views on mathematics as important progress when in 1944 he was asked to revise an entry about himself for a biographical dictionary and added the sentence: 'Wittgenstein's chief contribution has been in the philosophy of mathematics' (Monk 1990, 466). Accordingly, I shall focus on his most considered views, on his mature philosophy of the 1930s and 1940s, with only a brief sketch of his earlier views.

However, in spite of the relative importance that Wittgenstein accorded to his work on mathematics, he did not succeed in producing anything like a book or comprehensive and revised expression of his thoughts in that area. In the Preface (dated 1945) to *Philosophical Investigations* the reader is promised remarks on various topics, including the 'foundations of mathematics', but Wittgenstein changed his mind and decided to deal with the philosophy of mathematics separately, in another volume. The plan to write such a book was still there in 1949 (MS 169, 37), but was never carried out. All we have are one typescript and numerous unrevised manuscripts (or parts of manuscripts) with Wittgenstein's philosophical thoughts on mathematics, a collection of which was posthumously published as *Remarks on the Foundations of Mathematics*. It has even been doubted whether Wittgenstein's later philosophy of mathematics can be said 'to present something that amounts to a coherent view' (Potter 2011, 135). I shall try to show that that is an exaggeration. Although his thoughts never reached the stage of a fully worked-out and polished position, his general view is clear enough. At one point when his attention had shifted to issues in the philosophy of mind, Rush Rhees asked him what was to become of his work on mathematics and he replied lightly: 'Oh, someone else can do that' (Monk 1990, 466). Perhaps he himself thought that he had produced at least the rough outlines of a reasonable philosophical account of mathematics, which he could leave to others to flesh out in detail.

And yet there are undoubtedly considerable gaps, open questions, inconsistencies between different tentatively considered answers, and even tensions between some of the key ideas, in Wittgenstein's philosophy of mathematics. Therefore, drawing from his remarks a consistent position in the philosophy of mathematics is not just an exegetical challenge, but involves philosophical considerations. The reader has to decide which of a number of tentative or provisional remarks to put interpretative weight on and which ones to dismiss as carelessly phrased or as dead ends. Moreover, where the most promising and persuasive remarks remain very sketchy, it takes some extrapolation to present them as (part of) a plausible position. Therefore, what I present, though always following the general thrust of Wittgenstein's ideas, will occasionally go beyond them in its details. And since a plausible philosophical position is what I'm interested in, I shall not just present, but also discuss and evaluate Wittgenstein's ideas. By and large, I found myself able to defend Wittgenstein's views, mainly with arguments drawn from his own remarks, sometimes by recourse to additional considerations of my own. Sometimes I have remained unconvinced by his considerations and shall raise concerns or objections.

It appears that Wittgenstein's interest in philosophy was first properly awakened when mathematical problems that he encountered while studying engineering led him to reflect on the philosophy of mathematics. He was attracted by the foundationalist programme of logicism and in 1911 first visited Frege in Jena and then became a student of Russell's in Cambridge (von Wright 1954, 5–6). His account of mathematics in the *Tractatus* can be regarded as a form of 'logicism without classes' (Marion 1998, 21), while his subsequent thoughts on mathematics began, negatively, with his criticism and rejection of logicism. Hence, it seems appropriate to set the stage with a brief presentation of logicism and Wittgenstein's objections to it. He also came to reject other foundationalist positions in the philosophy of mathematics, such as formalism and intuitionism, but his critical discussion of them played a less important role in the development of his ideas and will be considered later, *after* the presentation of his positive views (formalism) or only in passing (intuitionism).

To complete the preliminary part of the book, I shall briefly sketch the development of his earlier ideas: first, his view of mathematics in the *Tractatus*; then, the ideas of his middle period in the early 1930s.

Part II tries to give a coherent, though by no means complete, picture of Wittgenstein's mature philosophy of mathematics. It is intended to cover all the main aspects of Wittgenstein's position, but certainly not all his discussions of philosophically interesting mathematical issues.

Just as the author of the *Philosophical Investigations* considers language as essentially embedded in human life and shaped by human needs and interests—replacing the *Tractatus* view of language as an abstract system of crystalline logical purity (*PI* §107)—so the later Wittgenstein regards mathematics as an 'anthropological phenomenon' (*RFM* 399e). For one thing, he emphasises that mathematics is a human activity—calculation—rather than a theory (*BT* 749). For another thing, he regards the results of such calculations as (akin to) grammatical propositions (*RFM* 162d): as norms for our use of language. These are the two key ideas of Wittgenstein's philosophy of mathematics: the *calculus view* and the *grammar view* of mathematics. The former dominates his thinking in the early 1930s, while the latter becomes his principal idea in the late 1930s and the 1940s, overshadowing the calculus view, but never replacing it entirely. The difficulty of giving a coherent account of Wittgenstein's later philosophy of mathematics is largely the difficulty of resolving the apparent tension between those two ideas.

Wittgenstein's anthropological approach to mathematics—regarding it as both an activity and a source of norms for other activities—obviously puts him in stark opposition to Platonism, the view that mathematical objects exist independently of us. This opposition Wittgenstein expresses in his slogan that 'the mathematician is an inventor, not a discoverer' (*RFM* I §168: 99). His position is much closer to formalism, but there, too, a crucial difference springs from Wittgenstein's emphasis of the anthropological dimension, in particular: his insistence that mathematics is not just a network of calculi, but essentially concepts and calculi produced with a view to (possible) application outside mathematics (*RFM* 257de).

Given Wittgenstein's conception of philosophy in general (see Schroeder 2006, 151–68), it should go without saying that he eschews any kind of revisionism. Philosophy, he says explicitly, 'leaves mathematics as it is' (*PI* §124), and again, in preliminary remarks to his lectures he declares that 'it will be most important not to interfere with the mathematicians' (*LFM* 13). He would disagree with mathematicians only where they advance dubious philosophical claims about their subject. It may appear though that Wittgenstein disregarded his own maxim of non-interference at least in one instance, namely in his scathing critique of set theory (see Rodych 2000; cf. Bouveresse 1988, 47–51). However, as he repeatedly explains, it is not the actual mathematics, the calculus of set theory that he finds fault with (*PG* 469–70, *LFM* 141), but its philosophical interpretation: the idea that it provides descriptions of actual infinite totalities (cf. Dawson 2016, 326–32). Thus, in a lecture, he comments on Hilbert's famous slogan 'No one is going to turn us out of the paradise [of transfinite set theory] which Cantor has created', by saying that he 'wouldn't dream of trying to drive anyone out of this paradise'—:

> I would try to do something quite different: I would try to show you that it is not a paradise—so that you'll leave of your own accord. I would say, "You're welcome to this; just look around you."
>
> (*LFM* 103)

According to Wittgenstein, there are no mathematical mistakes in set theory; it is just pointless: not really providing a foundation for other parts of mathematics and lacking any proper application (*RFM* 260a, 378bc). Yet he does not on that account deny set theory the status of mathematics (*pace* Rodych 2000, 305–8). For although he insisted that by and large mathematics needs extra-mathematical applications, he was well aware that not *every* mathematical calculus can be expected to find an application, and certainly not *immediately* (RR 132; *RFM* 399d).

I am grateful to Hanoch Ben-Yami, Patrick Farrell, Simon Friederich, Peter Hacker, Felix Hagenström, Wolfgang Kienzler, Felix Mühlhölzer, Antonio Scarafone, Harry Tomany, and Cosmin Vaduva for their comments on drafts of chapters or replies to my questions. Thanks to Brad Hooker for organising some funding for my work. Alois Pichler of the Wittgenstein Archive in Bergen deserves thanks for having made Wittgenstein's manuscripts easily accessible. I would also like to thank an anonymous reviewer for useful criticisms and suggestions. My greatest debts are to Kai Büttner and David Dolby, who read every part of the book and whose comments and discussions have been immensely helpful to me.

Parts of the book are based on previously published material of mine. Chapter 6 includes passages from 'Mathematical Propositions as Rules of Grammar' (2014). Chapter 7 draws on 'Intuition, Decision, Compulsion' (2016). Chapter 8 makes use of material from 'On some Standard Objections to Mathematical Conventionalism' (2017a) and 'Mathematics and Forms of Life' (2015). And some passages of Chapter 10.2 first appeared in 'Conjecture, Proof, and Sense in Wittgenstein's Philosophy of Mathematics' (2012).

Abbreviations

Ludwig Wittgenstein

AL — *Wittgenstein's Lectures, Cambridge, 1932–1935*, ed.: A. Ambrose, Oxford: Blackwell, 1979.

BB — *The Blue and Brown Books*, Oxford: Blackwell, 1958.

BT — *The Big Typescript: TS 213*, ed. & tr.: C.G. Luckhardt & M.A.E. Aue, Oxford: Blackwell, 2005.

CL — *Cambridge Letters: Correspondence with Russell, Keynes, Moore, Ramsey and Sraffa*, eds.: B. McGuinness & G.H. von Wright, Oxford: Blackwell, 1995.

LFM — *Wittgenstein's Lectures on the Foundations of Mathematics Cambridge, 1939*, ed.: C. Diamond, Hassocks, Sussex: Harvester Press, 1976.

LC — *Lectures and Conversations on Aesthetics, Psychology and Religious Belief*, ed.: C. Barrett, Oxford: Blackwell, 1978.

LL — *Wittgenstein's Lectures, Cambridge, 1930–1932*, ed.: D. Lee, Oxford: Blackwell, 1980.

LW I — *Last Writings on the Philosophy of Psychology. Volume I*, eds.: G.H. von Wright & H. Nyman, tr.: C.G. Luckhardt & M.A.E. Aue, Oxford: Blackwell, 1982.

MS — Manuscript in *Wittgenstein's Nachlass: The Bergen Electronic Edition*, Oxford: OUP, 2000.

NB — *Notebooks 1914–16*, ed.: G.H. von Wright & G.E.M. Anscombe, tr.: G.E.M. Anscombe, rev. ed. Oxford: Blackwell 1979.

PG — *Philosophical Grammar*, ed.: R. Rhees, tr.: A.J.P. Kenny, Oxford: Blackwell, 1974.

PI — *Philosophical Investigations*, eds.: P. M. S. Hacker & J. Schulte, tr.: G.E.M. Anscombe; P.M.S. Hacker, J. Schulte, Oxford: Wiley-Blackwell, 2009.

PLP — *The Principles of Linguistic Philosophy* by F. Waismann [based on dictations by Wittgenstein], London: Macmillan, 1965.

PPF — 'Philosophy of Psychology—a Fragment', in *PI* [= 'Part II' of previous editions].

PR	*Philosophical Remarks*, ed.: R. Rhees, tr.: R. Hargreaves & R. White, Oxford: Blackwell, 1975.
RFM	*Remarks on the Foundations of Mathematics*, eds.: G.H. von Wright, R. Rhees, G.E.M. Anscombe; tr.: G.E.M. Anscombe, rev. ed., Oxford: Blackwell, 1978.
RPP II	*Remarks on the Philosophy of Psychology*, Volume II, eds.: G.H. von Wright & H. Nyman; tr.: C.G. Luckhardt & M.A.E. Aue, Oxford: Blackwell, 1980.
RR	'On Continuity: Wittgenstein's Ideas, 1938' [notes of Wittgenstein's conversations taken by Rush Rhees], in: Rush Rhees, *Discussions of Wittgenstein*, London: Routledge, 1970; 104–57.
TLP	*Tractatus Logico-Philosophicus*, tr.: D.F. Pears & B.F. McGuinness, London: Routledge & Kegan Paul, 1961.
WVC	*Ludwig Wittgenstein and the Vienna Circle*. Conversations recorded by Friedrich Waismann, ed.: B. McGuinness, tr.: J. Schulte & B. McGuinness, Oxford: Blackwell 1979.
Z	*Zettel*, eds.: G.E.M. Anscombe & G.H. von Wright, tr.: G.E.M. Anscombe, Oxford: Blackwell, 1967.

A Note on Referencing *RFM*

Only the remarks of Part I of *Remarks on the Foundations of Mathematics* were ordered and numbered by Wittgenstein himself (in TS 222). The rest of the book is a selection and arrangement by the editors of remarks from various manuscripts. Their numbering style is somewhat different from Wittgenstein's own: as the sections often comprise a *series* of remarks (sometimes 3 or 4, sometimes 13 or 14), whereas Wittgenstein would give each remark a new number. Therefore, I cite only Part I by section number, followed by page number (e.g.: *RFM* I §162: 98), while references to other Parts are by page number and letter to specify the paragraph(s) (e.g. *RFM* 257de), which is both more convenient and more precise than references by part and section throughout.

Part I
Background

1 Foundations of Mathematics

Philosophy of mathematics has been to a large extent a discussion of the *foundations* of mathematics, that is, of attempts to show that all mathematical developments were reliably true, laying to rest any worries about unproven assumptions or hidden inconsistencies. Monographs on the philosophy of mathematics often contain three main sections covering the three major schools offering new and more rigorous foundations to mathematics: logicism, formalism, and intuitionism. At least at first glance, Wittgenstein's writings appear to be no exception to this trend. Of course, the title of his book *Remarks on the Foundations of Mathematics* was chosen by the editors, Anscombe, Rhees and von Wright, who are also responsible for the selection of remarks included in this posthumous collection. But that title is not inappropriate in light of some of Wittgenstein's own characterisations of his work (cf. Mühlhölzer 2010, 1–10). The Introduction to his *Philosophical Investigations* promises remarks on many subjects, including 'the foundations of mathematics' (which in the end he decided to remove with a view to a separate publication). And the first sentence of his 1939 lectures on the philosophy of mathematics is: 'I am proposing to talk about the foundations of mathematics' (*LFM* 13).

On the other hand, one of the main concerns of Wittgenstein's philosophy of mathematics was the rejection of any foundationalist project:

> What does mathematics need a foundation for? It no more needs one, I believe, than propositions about physical objects—or about sense impressions, need an *analysis*. What mathematical propositions do stand in need of is a clarification of their grammar, just as do those other propositions.
>
> The *mathematical* problems of what is called foundations are no more the foundation of mathematics for us than the painted rock is the support of a painted tower.
>
> (*RFM* 378bc)

If, in spite of his hostile attitude towards projects such as logicism, intuitionism, or formalism, Wittgenstein had no qualms describing his own work in the area as concerned with the foundations of mathematics, that is no doubt partly because he thought it important to provide a critical discussion of such foundationalist attempts, especially of logicism, having been a collaborator and friend of Bertrand Russell's, one of logicism's leading proponents. However, Wittgenstein often used the expression 'foundations of mathematics' in a different sense: for exactly the kind of 'clarification of the grammar' of mathematical expressions that would show up the foundational project as misguided.[1] Thus, in *The Big Typescript* he writes:

> what has to be done is to describe the calculus—say—of the cardinal numbers. That is, its rules must be given, and thereby the foundation is laid for arithmetic.
>
> Teach them to us, and then you have laid its foundation.
>
> (*BT* 540)

Again, in the final remark of the typescript that was published as 'Part II' of *Philosophical Investigations* he explains this different sense of the expression 'foundations' (*Grundlagen*):

> An investigation entirely analogous to our investigation of psychology is possible also for mathematics. It is just as little a *mathematical* investigation as ours is a psychological one. It will *not* contain calculations, so it is not for example formal logic. It might deserve the name of an investigation of the 'foundations of mathematics'.
>
> (*PPF* xiv)

The main part of this monograph will be devoted to a presentation and discussion of Wittgenstein's own account of the 'foundations of mathematics' in the latter sense of the expression: as a clarification of the role and function of mathematical language, the meaning and status of mathematical propositions. First, however, I shall turn to logicism, the foundationalist project that Wittgenstein was most familiar with, briefly characterise its motivation and give a brief outline of its basic ideas, before presenting and discussing Wittgenstein's objections to it. Wittgenstein's

1. Wittgenstein displayed the same kind of ambivalence with respect to the term 'meaning'. Taking the word in a referentialist sense (illustrated by Augustine's ideas) he rejects questions of meaning as irrelevant: 'No such thing [as meaning] was in question here, only how the word "five" is used' (*PI* §1). But later on he suggests a positive account of meaning as use (*PI* §43).

rejection of logicism, the position of his philosophical father figures, Frege and Russell, can be regarded as clearing the ground for the development of his own ideas. There are two other major contemporary ideas concerned with foundationalist endeavours that Wittgenstein discussed and criticised: the alleged importance of inconsistency proofs and Gödel's first Incompleteness Theorem. In those cases, however, Wittgenstein's response is very much a corollary of his positive account of mathematics, so I shall consider them only towards the end of my book.

***

Why should mathematics be thought to need a foundation? The motivation for a foundationalist project may be (i) epistemological or (ii) ontological. I shall briefly consider them in turn.

(i) *Epistemological concerns*. Morris Kline describes the history of mathematics as a tragedy: after a triumphant beginning of mathematizing the sciences in order to read the book of nature written in the language of mathematics, mathematicians experienced an increasing 'loss of certainty'. For since the development of analysis in the 17th century,

> mathematics had developed illogically. Its illogical development contained not only false proofs, slips in reasoning, and inadvertent mistakes which with more care could have been avoided. Such blunders there were aplenty. The illogical development also involved inadequate understanding of concepts, a failure to recognize all the principles of logic required, and an inadequate rigor of proof; that is, intuition, physical arguments, and appeal to geometrical diagrams had taken the place of logical arguments.
>
> (Kline 1980, 5)

Calculus was a powerful tool in physics, but its concepts were ill understood and apparently illogical. The idea that one could calculate instantaneous velocity, i.e. the speed of an object during an interval of time that is zero, while the object does not actually move, was quite incomprehensible.

Again, in the 18th century there arose the strange problem of infinite series. Consider the much disputed problem of computing the sum of the following infinite series of numbers:

[S] $\quad 1 - 1 + 1 - 1 + 1 - 1 \ldots$

On the one hand, it can be argued that *S* must equal 0, for it can be written thus:

$$(1 - 1) + (1 - 1) + (1 - 1) \ldots$$

I.e. as an infinite sum of bracketed expressions that all equal 0.

On the other hand, it can also be written as follows:

$$1 - (1 - 1) - (1 - 1) - (1 - 1) \ldots$$

(for $-1 + 1 = -(1 - 1)$). So the result must be 1, since one keeps subtracting 0 from 1.

But applying the same arithmetic rule according to which a minus sign in front of a bracket inverts minus and plus inside the brackets, S can also be re-written in this way:

$$1 - (1 - 1 + 1 - 1 + \ldots).$$

Hence:

$$S = 1 - S.$$

It follows that $S = ½$ (Kline 1980, 142; Giaquinto 2002, 6). It would appear from these three contradictory results that important parts of 18th-century mathematics contained inconsistencies.

In response to such conceptual puzzles and flagrant inconsistencies, mathematicians in the latter half of the 19th century made an effort to put mathematics on a clearly defined and rigorous footing. Foundationalist programmes grew out of that 19th-century rigorisation of mathematics. Thus, Russell describes his endeavours to reduce mathematics to logic as motivated by a craving for absolute certainty, of which traditional mathematics often fell short:

> I wanted certainty in the kind of way in which people want religious faith. I thought that certainty is more likely to be found in mathematics than elsewhere. But I discovered that many mathematical demonstrations, which my teachers expected me to accept, were full of fallacies.
>
> (Russell 1956, 53)

The logicist programme has two parts: (1) showing that all primitive concepts of mathematics can be defined in terms of purely logical terms; (2) showing that all axioms which serve as a basis for mathematics can be derived from a number of purely logical principles by demonstration. However, although a successful reduction of all mathematics to transparent concepts and rules of logic would indeed afford the desired certainty for mathematics, it is not a necessary condition for achieving that aim. To avoid the kind of conceptual confusion and inconsistency illustrated above, mathematicians have to give sufficiently precise and rigorous definitions of their fundamental concepts; but that does not mean that those concepts have to be reduced to purely logical concepts. Indeed, the desired rigorisation of the fundamentals of analysis—the area where conceptual problems and contradictions had appeared so problematic—was completed, mainly through the work of Karl Weierstrass (1815–97), by the end of the 19th century:

> Weierstrass's work finally freed analysis from all dependence upon motion, intuitive understanding, and geometric notions which were certainly suspect by Weierstrass's time.
>
> (Kline 1980, 177)

Yet such necessary clarification and rigorisation did not require or involve any logicist reductionism. In other words, Russell's concern for certainty makes it psychologically understandable that the logicist project appealed to him, but it provides no compelling reason to regard that project as necessary.

Still, it would appear that concerns about inconsistency justify the endeavour to provide a consistency proof for all areas of mathematics, and thus David Hilbert's foundationalist programme. To that I shall return in Chapter 11.

(ii) *Ontological concerns.* Gottlob Frege, the founding father of logicism, whose pertinent writings predate Russell and Whitehead's attempt by a couple of decades, was motivated by quite different considerations. Unlike Russell, he did not appear to worry about existing mathematics' lack of certainty. His primary concern was not epistemological, but ontological.[2] For the most part, he expressed no doubt about the *certainty* of the results of calculations or proofs, be it in elementary or advanced mathematics, but about our philosophical understanding of their *contents:* about the *real meaning* of mathematical propositions. Indignant at the philosophical carelessness with which mathematicians introduce their basic concepts, especially of numbers and operations, Frege writes:

> So [mathematical] science does not know what contents of thought to attribute to its theorems; it does not know what its subject matter is; it is entirely in the dark about its own nature. Is that not a scandal?
>
> (Frege 1889, 133)[3]

What scandalised Frege was that mathematicians had not been able to come up with plausible answers to the question 'What is a number?' understood to ask for the identification of particular kinds of objects: 'What objects do number words designate?'. He had no difficulties

2. Although occasionally he also expresses epistemological concerns: Frege 1884, xxi.
3. *Die Wissenschaft weiß also nicht, welchen Gedankeninhalt sie mit ihren Lehrsätzen verbindet; sie weiß nicht, womit sie sich beschäftigt; sie ist über ihr eigenes Wesen völlig im unklaren. Ist das nicht ein Skandal?*

showing up the implausibility of accounts regarding number words as names of either psychological or physical entities. And so he carefully developed his own much more sophisticated view of numbers as abstract objects, namely classes of equinumerous concepts. Thus, the sign '7', according to Frege, is the name of the class of all concepts that are instantiated by seven objects.

Two points deserve to be emphasised. First, Frege's motivation was philosophical, not mathematical. For the most part, he does not seem seriously concerned that the lack of proper foundations might engender mathematical problems or undermine the validity of mathematical research. Rather, he aimed for a more profound (i.e. philosophical) understanding of the nature (*Wesen*) of the contents of mathematical investigations.

Secondly, Frege's philosophy of mathematics is squarely based on his referentialist conception of language: the idea that words are meaningful in virtue of being names of something (objects or concepts). His naïve commitment to this idea is reflected in his very use of the word 'meaning' (*Bedeutung*) for: reference, the object denoted. Only on the assumption that meaning is (or is determined by) reference could it appear to Frege that mathematicians had not been able to offer any plausible answer to the question: 'What does the sign "7" mean?'. If, by contrast, one takes this question to be answered by an explanation of the arithmetical use of the sign '7' (cf. *PI* §1), it is obvious that any competent maths teacher could give at least a *roughly correct* answer (even though a precise and fully comprehensive *definition* may be more difficult to provide—just as in the case of other ordinary concepts).

2 Logicism

2.1 Frege's Logicism

Frege very plausibly analyses statements of number (—how many *F*s there are) as statements about a concept (1884, §46; cf. *RFM* 300c).[1] So the syntactic uniformity of the two following propositions is deceptive:

(1) Arsenal's first team are experienced players.
(2) Arsenal's first team are sixteen players.

Whereas the adjective 'experienced' characterises the objects denoted by the subject term, the word 'sixteen'—although it seems to occupy the same adjective position—does not characterise the objects denoted by the subject term, but the concept (of a member of Arsenal's first team). (1) says that every single member of Arsenal's first team is experienced, whereas (2) does not say that every single member of the team is sixteen (members). Rather, it says that the concept 'player in Arsenal's first team' applies to sixteen people (Frege 1884, §45).

This observation might be taken to suggest (though that was not ultimately Frege's view) that numbers function as second-order concepts: concepts to characterise other concepts, to say how many things fall under them. So they could plausibly be regarded as akin to the existential quantifier in formal logic, only more specific. Indeed, with the existential quantifier (and the universal quantifier defined in terms of the existential quantifier and negation),[2] number statements can easily be formalised:

(3)

$\sim\exists x\, Fx$	There are 0 *F*s.
$\exists x\, [Fx \,.\, (y)[Fy \supset x = y]]$	There is exactly 1 *F*.
$\exists x \exists y[Fx \,.\, Fy \,.\, x \neq y \,.\, (z)[Fz \supset [z = x \vee z = y]]]$	There are exactly 2 *F*s.
etc.	

1. For objections to Frege's analysis, see Rundle 1979, 254–71.
2. $(x)Fx =df \sim\exists x \sim Fx$

And in a next step, using that Fregean insight, we can define numeric quantifiers, using indices, accordingly:

(4)

$$\exists_1 xFx \quad =df \quad \exists x\, [Fx \,.\, (y)[Fy \supset x = y]]$$
$$\exists_2 xFx \quad =df \quad \exists x \exists y[Fx \,.\, Fy \,.\, x \neq y \,.\, (z)[Fz \supset [z = x \vee z = y]]]$$

etc.[3]

However, Frege is not satisfied with such an analysis of number words as numeric quantifiers, for although it explains attributive uses of number words, such as ‘There are 7 houses’ (or, as Frege puts it, ‘The number 7 belongs to the concept of a house’), it fails to give us a definition of numbers ‘as self-subsistent objects’ (Frege 1884, §56). Thus, we have an explanation of expressions such as ‘7 houses’, and we can develop the series of natural numbers, but we have no criterion to decide whether (to use Frege’s curious example) Julius Caesar is a number, too.

Frege is convinced that number words are proper names of self-subsistent objects for which we have to specify criteria of identity (1884, §62). After all, number words can also occur in subject position—e.g. ‘The number 163 is prime’ —, and they frequently occur in equations, which Frege interprets as identity statements about such objects: ‘*7 + 5 = 12*’ for example, means that *the object 7 + 5* is identical with *the object 12* (Frege 1884, §57).

Hence, attributive uses of number words, too, should be analysed as identity statements. Instead of ‘Jupiter has four moons’ it would be more appropriate to say (Frege 1884, §57):

(5) The number of Jupiter’s moons = the number four.

In other words, rather than saying that 4 is predicated of the concept of a moon of Jupiter’s, Frege claims that *being identical with the object 4* is predicated of the number belonging to the concept of a moon of Jupiter’s.

What kind of objects are numbers? They are extensions of concepts of equinumerous concepts (Frege 1884, §68):

(6) The number of *F*s *=df* the extension of the concept ‘equinumerous with concept *F*’.

3. Developing this kind of analysis further, one can then propose to express arithmetic propositions as logical truths, e.g. ‘*3 + 2 = 5*’ could be rewritten as follows:

(7) $[\exists_3 xFx \,.\, \exists_2 xGx \,.\, {\sim}\exists x[Fxs \,.\, Gx]] \supset \exists_5 x\, [Fx \vee Gx]$.

Wittgenstein attributes this kind of analysis of arithmetical equations to Russell (*WVC* 35, *LFM* 159, *RFM* 142–55), although it does not occur explicitly in *Principia Mathematica*, where instead Russell offers a corresponding set-theoretical analysis (cf. Marion 1998, 230: n. 24; Mühlhölzer 2010, 133–5).

In other words:

(8) The number of *F*s =*df* the class of all concepts equinumerous with the concept *F*.

For example, the number of seasons is the following class of concepts: *evangelist*, *member of the Beatles*, *cardinal point of the compass*, *wife of Bertrand Russell*, *suit in a deck of cards*, *crowned son of King Aethelwulf*, ... In short, the number two is the class of all concepts applying to two objects, the number three is the class of all concepts applying to three objects, the number four is the class of all concepts applying to four objects, and so on.

Having defined the expression 'the number of *F*s', Frege can then give the following definition of a cardinal number in general (Frege 1884, §72):

(9) *n* is a cardinal number (*Anzahl*) if and only if there is a concept *F* such that *n* = the number of *F*s.

But isn't definition (8) circular? It would appear that the concept of a number is defined in terms of 'equinumerous', which means: having the same number. Frege anticipates that objection and emphasises that 'equinumerous' is not to be explained in terms of number, but in terms of one-to-one correlation (1884, §63). After all, in order to check whether the plates and knives on a table are equinumerous, it is not necessary to count them: it suffices to ascertain that next to every plate is a knife and vice versa (Frege 1884, §70).

Next, Frege defines particular numbers beginning with *0*. Obviously, his logicist project requires that his definitions don't use any empirical concepts (whose extension is contingent), such as English kings or philosophers' wives. So Frege defines (Frege 1884, §74):

(10) *0* =*df* the number of the concept *not identical with itself*.

Then Frege defines the immediate successor relation between numbers as follows (1884, §76):

(11) *n* is an immediate successor of *m* if & only if *n* is the number of a concept *F*, and *x* is an object falling under concept *F*, and *m* is the number of the concept 'falling under *F* but not identical with *x*'.

Finally, he presents a way of defining the following natural numbers (Frege 1884, §77):

(12) *1* =*df* the number of the concept *identical with 0*.

(13) *2* =*df* the number of the concept *identical with 0 or 1*.
(14) *3* =*df* the number of the concept *identical with 0 or 1 or 2*.

and so on.

Given his assumption that numbers are objects, Frege requires an infinite supply of objects to define the series of natural numbers. However, these are abstract objects that can be seen, so to speak, automatically to generate their successors. Moreover, Frege is able to sketch a formal proof that the number series thus defined has no end (1884, §§82–3).

2.2 The Class Paradox and Russell's Theory of Types

Frege proposed to derive 'the simplest laws of cardinal numbers' from the definitions just presented and a particular system of what he regarded as logical axioms, among them his Basic Law V, which allows the transition from a concept to its extension (Frege 1893, 1; I §20: 36). In June 1902 Russell wrote to Frege pointing out a contradiction in his system. He asked him to consider the (apparent) predicate 'is a predicate which cannot be predicated of itself', and then to consider the corresponding idea about extensions or classes. As it appears licit to ask whether a given class is a member of itself, you can—according to Basic Law V—define (and thus establish the existence of):

(15) *R*: the class of all classes which are not members of themselves.

But if you consider whether this class, *R*, is a member of itself, you get the following contradiction, known as *Russell's Paradox*:

(16) *R* is a member of *R* if & only if *R* is *not* a member of *R*.
$(p \textit{ iff } {\sim}p)$[4]

Thus, Frege's axiomatic system which was meant to give a firm logical foundation to arithmetic is flawed, as it allows the proof of a contradiction (and from a contradiction you can deduce anything whatsoever). It was imperative to find a way of avoiding Russell's Paradox, or the programme of providing a logical foundation to arithmetic had to be given up.

4. By way of illustration, suppose that a village barber shaves all and only the men in the village who do not shave themselves. Does the barber shave himself? If he does, it follows that he is a man who does not shave himself; but if he does not shave himself, it follows that he does! However, as Russell pointed out, this is only an approximate illustration (Russell 1918, 261), for, unlike the original paradox, it allows the easy solution that such a barber must be a woman or doesn't exist, whereas the existence of a correctly defined class cannot be denied.

It appears that as Wittgenstein was drawn into engineering by the challenge of constructing and flying an aeroplane, he was drawn into philosophy by the challenge of finding a solution to Russell's Paradox. Already in 1909, when still experimenting with his propeller engine, he had had a go at it and sent his attempted, yet unsuccessful, solution to the mathematician Philip Jourdain, a friend of Russell's (McGuinness 1988, 74–6).

Russell himself, in *Principia Mathematica* (1910–13), proposed to avoid his paradox by a Theory of Types. Objects and classes were to be ordered in a hierarchy of types:

Type 0: individual objects;
Type 1: classes of objects;
Type 2: classes of type 1 classes;
Type 3: classes of type 2 classes; etc.

And the formation of classes was to be restricted such that the members of a class must always be of a lower type than the class itself. Thus the possibility of a class's being a member of itself is ruled out, and hence Russell's Paradox can no longer arise.

However, given this restriction on the formation of classes, when Russell proceeded to offer his own definition of numbers in terms of classes, he could not define them as classes of *all* equinumerous classes, regardless of the type of their members (corresponding to Frege's definition of numbers as classes of equinumerous concepts). Instead, Russell defined numbers as classes of equinumerous classes of *objects* in the narrow 'type 0' sense defined by his Theory of Types, excluding classes. That meant, however, that an infinity of numbers required an infinity of (type 0) empirical objects. So Russell had to rely on an *Axiom of Infinity*, stating that the number of (type 0) objects in the world was not finite. Wittgenstein had good reason to be dissatisfied with this solution, for even if the Axiom of Infinity were true, it would certainly not be a truth of logic, and so its inclusion in *Principia Mathematica* defeated the book's ambition of providing a purely logical foundation to arithmetic.[5]

2.3 *Tractatus Logico-Philosophicus*: Logicism Without Classes

Wittgenstein thought that Russell's Theory of Types should be 'rendered superfluous by a proper theory of the symbolism' (*CL* 24); but more dramatically, while holding on to the logicist view of mathematics as 'a method of logic' (*TLP* 6.234), he was going to reject the Frege-Russell approach of taking the theory of classes as the foundation of arithmetic

5. We shall return to this criticism in Chapter 3.

(*TLP* 6.031). In the account of arithmetic given in his first book, the *Tractatus Logico-Philosophicus*, the notion of a class is replaced by that of a *logical operation*.[6] Numbers are construed as exponents of operations, i.e. the numbers of applications of a symbolic procedure (*TLP* 6.02f.).[7] However, the logical reconstruction of mathematics is given very little space in the *Tractatus*. It seems that when Wittgenstein arrived in Cambridge his main interest had already shifted from the logical nature of mathematics to the nature of logic itself.

6. Rodych (2018, §1) rightly points out that the *Tractatus* account of mathematics is not logicism in Frege's or Russell's sense, 'because Wittgenstein does not define numbers "logically" in either Frege's way or Russell's way and the similarity (or analogy) between tautologies and true mathematical equations is neither an identity nor a relation of reducibility', but it still seems natural to label it as a (non-reductive) kind of logicism (Frascolla 1994, 37; Marion 1998, Ch.2; cf. Chapter 4.1).
7. For a general account of Wittgenstein's philosophy of mathematics in the *Tractatus*, see Frascolla 2001, Ch.1.

3 Wittgenstein's Critique of Logicism

Wittgenstein's writings, conversations, and lectures contain six major objections to (reductive) logicism, which I shall consider in turn.

3.1 Can Equality of Number Be Defined in Terms of One-to-One Correlation?

Remember the key idea in Frege's (and Russell's) definition of number:

> (1) The number of *F*s = the number of *G*s *if & only if* there exists a relation ϕ of one-to-one correspondence between the *F*s and the *G*s.

In 1931 in conversation with members of the Vienna Circle, Wittgenstein presented the following objection to that step:

> Imagine I have a dozen cups. Now I wish to tell you that I have got just as many spoons. How can I do it?
>
> If I wanted to say that I allotted one spoon to each cup, I would not have expressed what I meant by saying that I have just as many spoons as cups. Thus it will be better for me to say, I can allot the spoons to the cups. What does the word 'can' mean here? If I meant it in the physical sense, that is to say, if I meant that I have the physical strength to distribute the spoons among the cups—then you would tell me, We already knew that you were able to do that. What I mean is obviously this: I can allot the spoons to the cups because there is the right number of spoons. But to explain this I must presuppose the concept of number. It is not the case that a correlation defines number; rather, number makes a correlation possible. This is why you cannot explain number by means of correlation (equinumerosity). You must not explain number by means of correlation; you can explain it by means of possible correlation, and this precisely presupposes number.
>
> You cannot rest the concept of number upon correlation.
>
> . . .

> When Frege and Russell attempt to define number through correlation, the following has to be said:
>
> A correlation only obtains if it has been *produced*. Frege thought that if two sets have equally many members, then there is already a correlation too. . . . Nothing of the sort! A correlation is there only when I actually correlate the sets, i.e. as soon as I specify a definitive relation. But if in this whole chain of reasoning the *possibility* of correlation is meant, then it presupposes precisely the concept of number. Thus there is nothing at all to be gained by the attempt to base number on correlation.
>
> (*WVC* 164–5)

In the case of cups and spoons and the relation of *being next to* it is obviously not correct to say that equinumerosity consists in the existence of that relation. That would mean that cups and spoons are of the same number only as long as they are thus laid out on the table, and the moment they are put back into their respective cupboards they cease being equinumerous. No, what we mean by equinumerosity is obviously independent of where and how the objects are arranged. So if spatial correlation is what is meant, then we cannot ask for *actual* correlation, but only for *possible* correlation.

At this point, however, Frege and Russell would demur, saying that it is not necessarily a *spatial* or physical correlation (like being placed next to) that they had in mind. Yes, they do speak of an actual correlation—of an existing relation ϕ that correlates both sets (Frege 1884, §72; cf. *LFM* 161)—but for them it suffices to have a relation that holds simply in virtue of the properties of the things involved (such as *having the same weight*, for example). But in the case of cups and spoons what kind of already existing correlation could that be? Perhaps the crockery and cutlery in a given household is personalized, such that each spoon and each cup bears the name of a different family member. Then, indeed, we can say that cups and spoons in that household are equinumerous if and only if they are one-to-one correlated by the relation *bears the same name tag*. But that is obviously something we cannot assume to be true in all cases. Often we are concerned with the number of things that have no individualizing features, but are virtually indistinguishable, like the unblemished spoons of a set. Russell tries to overcome that problem, Wittgenstein notes, 'as Frege did, by the relation of identity' (*LFM* 161):

> There is one relation which holds between any two things, *a* and *b*, and between them only, namely the relation $x = a \,.\, y = b$. (If you substitute anything except *a* for *x* or anything except *b* for *y* the equations become false and so the logical product is false.) You go on for 2-classes:
>
> $$\begin{matrix} a\,b & & c\,d \\ x = a \,.\, y = c & .\vee. & x = b \,.\, y = d \end{matrix}$$

> And so you go on to classes of any number. And so we get to the surprising fact that all classes of equal number are already correlated one-one.
>
> (*LFM* 162; cf. *PG* 356)

Russell held that you can always turn a list into a property—namely the property of belonging to that list. Thus, he writes, the class consisting of three men Brown, Jones, and Robinson can be defined by a common property possessed by them and 'by nothing else in the whole universe, namely, the property of being either Brown or Jones or Robinson' (Russell 1919, 12–13). Similarly, according to Russell, you can turn any possible one-to-one correlation between the members of two classes into an existing relation. Even if you despair of finding any similarity (or dissimilarity) peculiar to, say Tom and Mary, you can always single them out as the only pair of objects standing in the relation *x* = *Tom* . *y* = *Mary*. And if you need a relation that exclusively one-to-one correlates two classes, say the Beatles and Aethelwulf's crowned sons, you just one-to-one correlate their respective names in disjuncts of the same form, namely:

> *x* = *John* . *y* = *Aethelbald* .∨. *x* = *Paul* . *y* = *Aethelbert* .∨. *x* = *George* . *y* = *Aethelred I* .∨. *x* = *Ringo* . *y* = *Alfred the Great.*

This is a relation (i.e. a two-place predicate *Rxy*) that applies only to those four pairs:

> John, Aethelbald
> Paul, Aethelbert
> George, Aethelred I
> Ringo, Alfred the Great

Thus it is a relation that provides a one-to-one correlation between the Beatles and Aethelwulf's crowned sons.

It seems that in 1931 Wittgenstein objected to this procedure on the grounds that he had misgivings, explained in the *Tractatus* (*TLP* 5.53ff.), about the use of the identity sign (*WVC* 165). In 1939 he objected, more effectively, that it is rather misleading to call the possibility of correlating names in this manner the 'existence of a relation'. What it boils down to is that John is John and Aethelbald is Aethelbald. Does that really mean that there exists a relation between the two? (*LFM* 163) Clearly, the word 'relation' is used here in an artificially extended sense. That in itself may not be objectionable, but the important point to remember is that one-to-one correlation was introduced by Frege as a *criterion of identity* for numbers regarded as objects, that is, as a way of telling whether *x* and *y* are the same number. And as Kai Büttner observes, that there *exists* an 'extensive relation' of the kind envisaged by Russell really just means that it is *possible* to produce a one-to-one correlation between two lists of names (Büttner 2016b, 161–2). So effectively we can agree with Wittgenstein that in such cases (where no qualitative or spatial correlation can be found): 'A correlation only obtains if it has been *produced*' (*WVC* 165).

Then the next question is: What does it mean to say that it is *possible* to correlate cups and spoons? Wittgenstein suggests three answers. First, it could mean 'that I have the physical strength to distribute the spoons among the cups'. That goes without saying, Wittgenstein retorts. More seriously, Büttner adds, it would make number statements subjective (Büttner 2016b, 153), though that could be mended by making it a general statement about what is *humanly* possible. But what would be excluded by speaking of what is humanly possible? Well, in the way in which we can correlate cups and spoons (by laying them side by side) we could not correlate cups and stars (Waismann 1982, 45). But Russell's so-called extensive relation shows a way out: when dealing with unwieldy, non-spatial, historical or otherwise inaccessible objects we give them names, and then correlate tokens of those names. In any case, physical power is evidently not what we are interested in here.

So, secondly, Wittgenstein suggests a more pertinent reading of what it means to say that it is *possible* to effect a one-to-one correlation: 'I can allot the spoons to the cups because there is the right number of spoons.' And proceeds to object: 'But to explain this I must presuppose the concept of number' (*WVC* 164).—Indeed, if Frege tries to define sameness of number in terms of possible one-to-one correlation, he better not explain possibility of one-to-one correlation in terms of sameness of number—or the whole project would be circular.

A little earlier Wittgenstein had put his objection like this:

> In Russell's theory only an *actual* correlation can show the 'similarity' of two classes. Not the *possibility* of correlation, for this consists precisely in the numerical equality.
>
> (*PR* 140)

However, this second sentence Frege and Russell could accept. Possibility of one-to-one correlation is meant to be a necessary and sufficient condition of numerical equality. So of course it is true to say that spoons and cups can be one-to-one correlated because they are equal in number, or vice versa.[1] It is just that for the purpose of Frege's definition we must not *explain* one-to-one correlation in terms of what *it* is meant to explain.

1. Similarly, there is a species opposition in the wording from *WVC* 164: 'It is not the case that a correlation defines number; rather, number makes a correlation possible'. Of course correlation is made possible by [sameness of] number; but that is *why* one might plausibly propose to define same number in terms of possible correlation.

Can we not find another—non-circular—explanation of the required possibility of one-to-one correlation?

A third suggestion made by Wittgenstein (and following him by Waismann) is that 'it is *possible* to correlate cups and spoons' means that it *makes sense to say* that they are one-to-one correlated (*PR* 141b; Waismann 1936, 104; 1982, 50). This relates back to one of the main doctrines underlying the *Tractatus*, namely, the Bipolarity principle. According to that principle, for a proposition to be meaningful it must be contingent: it must be possible for it to be true and possible for it to be false. While in the *Tractatus* the principle was to explain meaningfulness, here it appears to be applied in the opposite direction: as an explanation of possibility.

However, on that interpretation of 'possible' too many correlations turn out to be possible. For whenever two classes are identified by empirical concepts (rather than by lists) it is a contingent matter whether the same number of objects fall under both concepts.[2] For example, there could be as many cats as horses, or there could be more or fewer—it's a contingent matter. Hence, according to the Bipolarity principle, the proposition 'Cats and horses are one-to-one correlated' is meaningful—it could be true or it could be false. So in this sense of 'possible', it is possible to one-to-one correlate cats and horses—i.e. meaningful to say that they are one-to-one correlated—even if, as it happens, there are a lot more cats than horses (so that the correlation claim happens to be false) (Waismann 1936, 104).

However, as Büttner points out (2016b, 158), there is yet another way in which one can explain the statement 'It is possible to produce a one-to-one correlation between cups and spoons'. The task is to specify the circumstances under which it is possible to achieve a one-to-one correlation—without using the concept of number (so as not to make the overall explanation of number circular). And that can be done by a procedural (counterfactual) conditional:

> 'It is possible to produce a one-to-one correlation between *F*s and *G*s' *means:* 'If one were to one-one correlate (be it physically by juxtaposing them or symbolically by juxtaposing their names) either all *F*s with *G*s or all *G*s with *F*s, there would be no *G*s or *F*s left uncorrelated.'

In other words, by the word 'possible' we don't want to rule out physical weakness; we want to rule out the case where there are more objects of one class than of the other—yet without using the concept of number. So instead, we describe the case to be ruled out as one where if the maximum

2. Unless, of course, they are inconsistent or contain number determiners.

correlating is done there is something left over and uncorrelated in one of the classes.

Given that explanation of possibility of correlation, there is no circularity in Frege's explanation of sameness of number (and ultimately number) in terms of possible one-to-one correlation.

Boudewijn de Bruin argues, on the contrary, that Wittgenstein's objection succeeds in showing up a circularity in Frege's procedure. His crucial claim is that if I haven't actually correlated *F*s and *G*s, I can only *know* that they can be correlated by knowing their numbers (de Bruin 2008, 365). That is true, but doesn't help Wittgenstein's argument and, as far as I can see, poses no problem to Frege's explanatory strategy (cf. Büttner 2016b, 172–3). After all, in order to understand the concept of possible correlation I don't have to know in any given case whether it applies. At most I have to know how to find out, and that I can do (as explained) without knowing the numbers: by effecting a one-to-one correlation.

It is true that at one point in the 1939 lectures Wittgenstein suggests, what perhaps de Bruin has in mind, that one-to-one correlation may not always be possible:

> At first you thought of cases where the correlation was the criterion. But if the correlation isn't possible, then it is the other way round: if they have the same number by such-and-such a criterion, then it is possible for them to be correlated.
>
> (*LFM* 158)

True, but why should correlation not be possible?

Marion and Okada (2014) argue that Wittgenstein's objection to the use of the notion of one-to-one correlation in the Frege-Russell definition of number (which they call the 'modality argument') is closely related to his 'surveyability argument' (see Chapter 3.5). They claim that all the criteria for producing a one-one correlation 'peter out when numbers grow large enough' (2014, 72). For example:

> one would not be able to correlate with any amount of certainty two sets of 3 million elements by drawing lines. . . . [Hence, eventually] one is left with no other choice but to count.
>
> (Marion & Okada 2014, 72)

However, Marion and Okada consider only three procedures of one-to-one correlation: subitising, arranging sets in familiar patterns, and drawing lines. Yet as already noted, there is also the possibility of indirect correlation by means of names. It is not obvious why that should not be feasible even with considerably larger sets. Of course, in the mediate correlation of sets of *millions* of objects errors could easily occur, but then, as Büttner points out, with such very large numbers counting becomes unreliable, too (2016b, 172).

Indeed, as Büttner remarks, counting two sets is itself a method of mediate correlation (2016b, 172): you correlate one of the *F*s and one of the *G*s by associating them both with the same counting word. And wherever you can do that, you could obviously also effect a correlation that was not yet counting: by choosing some other sequence of signs as intermediary correlates. Thus you may correlate the chimes of a church bell with the words of Andrew Marvell's poem 'To his Coy Mistress' and later do the same with the sheep in a field. If on both occasions you get to the word 'crime' (assuming it occurs only once), you have shown chimes and sheep to be equinumerous.

Could Wittgenstein reply that this would, in effect, be a form of counting, using the words of the poem as our numbers? Well, once this kind of procedure became common practice and we learnt to tell immediately which word comes before or after which other words—e.g. that 'enough' was fewer than 'this', but more than 'but'—then the words of the poem might indeed be regarded as our number series. But as long as no particular poem has been conventionally accepted as a counting device, to say that a flock of sheep can be correlated with the words up to 'crime' in 'To his Coy Mistress' amounts to nothing more than giving them names and writing them out in a list. Yet a list of 13 words is not the number 13.[3]

In conclusion, Wittgenstein's first objection to logicism fails. Since the possibility of one-to-one correlation can be explained without recourse to numbers, a definition of equinumerosity in terms of possible one-to-one correlation need not be circular.

3.2 Frege's (and Russell's) Definition of Numbers as Equivalence Classes Is Not Constructive: It Doesn't Provide a Method of Identifying Numbers

In Friedrich Waismann's notes of Wittgenstein's explanations we also find the following critical discussion of the logicist definition of number:

> *Definitions are signposts. They show the way towards verification.*
> ...

3. Another concern Wittgenstein expresses about the Frege-Russell definition of number based on one-to-one correlation is that it doesn't tell us what in a given case is to count as one-to-one correlation (*LFM* 156–60; cf. Bangu 2016, §3). That is not so much an objection to bringing in the procedure of one-to-one correlation (which in that respect is not more problematic than counting itself) as a suggestion for what could more properly be called a *foundation* of arithmetic: the practical skill of correlating or counting objects (just as geometry could be said to be based on certain techniques of measuring lengths (*LFM* 158)).

> According to the Frege-Russell principle of abstraction the number 3 is the class of all triples. Here we have to ask: does this definition show us the way towards verification?
>
> Do we verify the proposition: 'Here are 3 chairs' by comparing the class of these chairs with all other triples in the world? No! But if we can understand the sense of that proposition without verifying it in such a way then the proposition *itself* must already contain everything that is essential and reference to those triples cannot be relevant to the number 3.
>
> If I ask: 'How many chairs are in this room?' and I receive the answer: 'As many as in that room', I would rightly say: 'That is not an answer to my question. I asked how many chairs are here, and not where else I can find the same amount'.
>
> Russell's definition fails to achieve exactly what matters. Giving the number must contain a method for getting to that number. And that is exactly what that definition lacks.
>
> (*WVC* 221)

Wittgenstein's point is that there is a conceptual link between our concept of number and our technique of counting. A natural number is essentially the result of a count. That's how you get to the number *3*: by counting *1*, *2*, *3*. In other words, the number *3* is defined by its position in third place of the series of counting numbers.

Imagine a tribe of people that cannot count, but have a word translatable as: 'correlatable with the fingers of a hand'. Wittgenstein's point would be that such a word would not have the same meaning as our word '*5*', which means: successor of *4*—an expression that those people lack. Their expression is semantically poorer than our number word.

3.3 Platonism

A possible response on Frege's behalf might be that even though his definition of numbers in general is not constructive, it is to be seen together with his subsequent definition of particular numbers.

(8) The number of *F*s =*df* the class of all concepts equinumerous with the concept *F*.
(10) *0* =*df* the number of the concept *not identical with itself*.
(12) *1* =*df* the number of the concept *identical with 0*.
(13) *2* =*df* the number of the concept *identical with 0 or 1*.
(14) *3* =*df* the number of the concept *identical with 0 or 1 or 2*.

(8) and (14) together yield:

(17) *3* = the class of all concepts equivalent to the concept *identical with 0 or 1 or 2*.

So, although Frege's definition of number fails to capture our concept of a number, his list of definitions of specific numbers can be seen to afford the 'signpost towards verification' that Wittgenstein demands. For it shows us how to count objects by correlating them with the abstract objects *0, 1, 2, 3*. . . . Thus, the number *3* appears to be defined—somewhat oddly—as the result of counting to *2*, having started with *0*. And this oddity is of course the result of his Platonism: of his insistence that numbers be self-subsistent objects, rather than symbolic devices for counting.

There is something bizarre about Frege's (and Russell's) identification of numbers with classes of equinumerous concepts (or classes of equinumerous classes) in that it gives number words a reference that most competent speakers would find surprising. You can be an accomplished arithmetician, thoroughly familiar with numbers, and yet never have thought of classes (or sets) of equinumerous concepts (or classes). The Cambridge mathematician Timothy Gowers remarked that the claim that numbers 'really are' special kinds of sets 'is, of course, ridiculous, and probably almost nobody, when pressed, would say that they actually believed it' (Gowers 2006, 189). The step from a number statement to Frege's analysis of such a statement is comparable to the following pair:

(18) I like kale.
(19) I am a member of the set of all kale lovers.

It is true that (18) and (19) are logically equivalent, and so, once you introduce the concept of a set, you can derive (19) from (18). However, (18) itself does not involve the concept of a set (or class). Discovering and expressing your liking for kale doesn't require that you think of others: that you see yourself as one of an indefinitely large set of people sharing that predilection.

Similarly, and contrary to Frege's analysis, ascertaining that there are seven apples on the table by counting them does not involve any idea of comparison to all other classes of seven objects in the universe. Hence, the idea that the class of those countless classes is the object denoted by the word 'seven' appears highly artificial.[4]

Indeed, the very idea that number words should be the names of objects is questionable, as Wittgenstein shows at the very beginning of *Philosophical Investigations:*

> Now think of the following use of language: I send someone shopping. I give him a slip marked "five red apples". He takes the slip to

4. As first pointed out by Paul Benacerraf (1965), and again by Intisar-ul-Haque, over the 20th century various different, non-equivalent set-theoretic definitions of number have been developed, all of them satisfactory instantiations of Peano's axioms. Yet: 'If all these diverse, rather conflicting definitions are equally good then it is obvious that the definition of numbers is neither important nor obligatory' (Intisar-ul-Haque 1978, 55). None of them can plausibly be claimed to be necessary to understand the meaning of number terms.

> the shopkeeper, who opens the drawer marked "apples"; then he looks up the word "red" in a table and finds a colour sample opposite it; then he says the series of cardinal numbers—I assume that he knows them by heart—up to the word "five" and for each number he takes an apple of the same colour as the sample out of the drawer. . . . —But what is the meaning of the word "five"?—No such thing was in question here, only how the word "five" is used.
>
> (*PI* §1)

As the following sentence makes clear, the concept of meaning that Wittgenstein rejects here as irrelevant is the 'philosophical concept of meaning' (*PI* §2), introduced earlier through a quotation from Augustine: the meaning that 'is correlated with the word' as 'the object for which the word stands' (*PI* §1b). The object for which the number word 'five' stands—? 'No such thing was in question here, only how the word "five" is used'.

This is not a denial of the possible existence of abstract objects. Believe in the existence of a Platonic Form *Five* if you like. The point is merely that no such philosophical construct has any function in the actual use of language, at least at this everyday level. Platonism about numbers is just a fanciful philosophers' pastime, a pointless mythological embellishment.

But don't we say that number words stand for numbers and that a mathematical proposition, such as '*20 + 15 = 35*' is a statement about numbers (*LFM* 112)? Yes, we can use such formulations, but Wittgenstein explains (with regard to a simplified language game that contains words for building materials, number words, and indexicals) why it would be a mistake to regard that as ontologically significant:

> Now what do the words of this language *signify*?—What is supposed to shew what they signify, if not the kind of use they have? And we have already described that. So we are asking for the expression "This word signifies *this*" to be made a part of the description. In other words the description ought to take the form: "The word . . . signifies. . .".
>
> Of course, one can reduce the description of the use of the word "slab" to the statement that this word signifies this object. This will be done when, for example, it is merely a matter of removing the mistaken idea that the word "slab" refers to the shape of building-stone that we in fact call a "block"—but the kind of '*referring*' this is, that is to say the use of these words for the rest, is already known.
>
> Equally one can say that the signs "a", "b", etc. signify numbers; when for example this removes the mistaken idea that "a", "b", "c", play the part actually played in language by "block", "slab", "pillar". And one can also say that "c" means this number and not that one;

> when for example this serves to explain that the letters are to be used in the order a, b, c, d, etc. and not in the order a, b, d, c.
>
> But assimilating the descriptions of the uses of words in this way cannot make the uses themselves any more like one another. For, as we see, they are absolutely unlike.
>
> (*PI* §10)

Schemas of semantic explanation have a tendency to make linguistic meaning appear more uniform than it is, and in particular, they tend to assimilate everything to the word reference model of explanation. But in the case of number words, all that an explanation of the form 'This word signifies [*bezeichnet*] that number' amounts to is, first, that it is a word with a certain use, namely counting objects, and, secondly, that it has a certain position in the series of number words (e.g. between '*3*' and '*5*').

Earlier I suggested that in light of Frege's valuable insight that number statements say something about concepts it would be natural and plausible to regard numbers as (more specific) kinds of quantifiers. That is, they are comparable not only to the two quantifiers of formal logic, but also to ordinary language quantifiers, with their (often practical) vagueness, such as: 'many', 'few', 'the majority of', 'most', 'hardly any', etc. They, too, are plausibly construed as second-order concepts. Yet who would be inclined to construe 'many' or 'hardly any' as names of objects (cf. Benacerraf 1965, 60)?

I also mentioned that Frege's main reason for regarding numbers as self-subsisting objects was that in mathematics they often occur in subject position in a sentence and, especially, in equations, which appear to be statements about the identity of an object. However, such syntactic features are an unreliable guide to semantics. For example, there is no good reason to construe the subject term of 'It is raining' as referring to an agent. And the subject term of 'The whale is a mammal' is not a proper name (Intisar-ul-Haque 1978, 52). Moreover, there is no need to regard equations as identity statements; they can be plausibly understood as substitution rules, as Wittgenstein suggested in the *Tractatus* (*TLP* 6.23); comparable to:

> the majority of = more than half of
> often = frequently.

The alternative to Frege's Platonism, which Wittgenstein is going to develop in his later philosophy, is the view that 'arithmetic does not talk about numbers, it works with numbers' (*PR* 130). Number words, in whatever syntactic position, can be explained as linguistic devices that we use in our descriptions of the world without being cast in the role of names of objects to be described.

3.4 Russell's Reconstructions of False Equations Are Not Contradictions

As was briefly mentioned earlier, in Russell's logicist project equations are reconstructed as logical truths. Thus, '*3 + 2 = 5*' is analysed as follows:

(5) $[\exists_3 x Fx \,.\, \exists_2 x Gx \,.\, \sim\exists x[Fx \,.\, Gx]] \supset \exists_5 x\ [Fx \vee Gx]$.

In order to ascertain that *3 + 2 = 5*, we would have to show that (5) is a logical truth. However, in one of his 1939 lectures, Wittgenstein points out a telling difference in meaning between the equation and the logical truth. If a correct arithmetical equation comes out as a logical truth, then an incorrect equation should be a logical falsehood: a contradiction. The arithmetical proposition that *4 + 3 = 9* is not just contingently false, but inconsistent. However, the corresponding logical formula:

(20) $[\exists_4 x Fx \,.\, \exists_3 x Gx \,.\, \sim\exists x[Fx \,.\, Gx]] \supset \exists_9 x\ [Fx \vee Gx]$

is a contingent proposition (*LFM* 284). As Wittgenstein explains:

> Take '$\exists_4 x Fx \supset \exists_5 x Fx$'. Is this a contradiction? Of course not. "(Four people are in this room) ⊃ (five people are)" says "~(Four people are in this room) ∨ (five people are)". '$p \supset \sim p$' is not contradictory. It simply says $\sim p \vee \sim p$, which equals $\sim p$.
>
> (*LFM* 284; notation changed)

Thus, (20) means:

(21) If there are exactly 4 *F*s and 3 *G*s, and no *FG*s, then there are exactly 9 *F* ∨ *G*s.

But since the consequent is inconsistent with the antecedent, (21) implies the negation of its antecedent:

(22) It is not the case that there are exactly 4 *F*s and 3 *G*s, and no *FG*s.

Taking '*F*' and '*G*' to be replaceable by any predicates, that would amount to the claim that there are no two properties such that one is had by exactly four objects and the other by exactly three different objects. If that is an empirical falsehood, it is out of sync with the corresponding arithmetical equation, which is necessarily false.

This is related to one of the standard objections to *Principia Mathematica* (already mentioned earlier). In order to avoid the inconsistencies that could be derived from Frege's concept of a class (or extension) central to his definitions of numbers, Russell introduced his Theory of Types, proscribing the formation of multi-level classes. That led him in his, otherwise similar, definition of numbers to rely on classes of classes of *objects* (which must not include classes). Consequently, he required the Axiom of Infinity to assure us that the universe contains an infinite supply of objects in terms of which to define the infinity of natural numbers. Without the Axiom of Infinity we might come across a natural number that has no successor. As Russell himself acknowledged: 'we should be left with the possibility that [for some *n*] *n* and *n* + *1* might both be the null-class' (Russell 1919, 132). This evidently puts paid to the logicist programme of reducing arithmetic to pure logic. The Axiom of Infinity is not a logical truth (cf. *TLP* 5.535), indeed, it is not even known to be a truth: it is a verification-transcendent cosmological claim. This amounts not only to an impurity of the proposed reduction (since more is involved than pure logic); it means that the reduction itself fails. For whereas it is an essential definitional feature of our system of natural numbers that each number has a unique immediate successor, that could only be contingently true of the numbers defined by Russell. Hence, his concept of number is markedly different from the arithmetical one in its logical content. As a consequence, the epistemological standing of the two calculi is very different: whereas arithmetic is certain, Russell's reconstruction may well lead to falsehood, if the Axiom of Infinity happens to be false (Körner 1960, 59–60; Waismann 1982, 51–4).

3.5 Frege's and Russell's Formalisation of Sums as Logical Truths Cannot Be Foundational as It Presupposes Arithmetic

Wittgenstein's main objection to Russell's attempt to analyse equations as logical truth is this:

> The correctness of an arithmetical proposition is never expressed by a proposition's being a tautology. In the Russellian way of expressing it, the proposition 3 + 4 = 7 for example can be represented in the following manner:
>
> $(\exists_3 x)\varphi x \,.\, (\exists_4 x)\psi x \,.\, \sim(\exists x)\varphi x \,.\, \psi x : \supset : (\exists_7 x).\ \varphi x \vee \psi x.$
>
> Now one might think that the proof of this equation consisted in this: that the proposition written down was a tautology. But in order to be able to write down this proposition, I have to *know* that 3 + 4 = 7. The whole tautology is an application and not a proof of arithmetic.
>
> (*WVC* 35; cf. 106)

> The calculation holds if a certain expression is tautological. But whether it *is* tautological presupposes a calculation.—In the case of a thousand terms it would be by no means obvious.
>
> (*LFM* 159)

Wittgenstein's use of the word 'tautology' in these passages is a little odd. The word is obviously not to be taken in its standard modern sense since the formulae in question are not of the propositional calculus. What he means must be: logical truth, or formula derivable in Russell's calculus (Mühlhölzer 2010, 134–6).

Wittgenstein denies that one could prove $3 + 4 = 7$ by proving that the logical formula '$(\exists_3 x)\varphi x \,.\, (\exists_4 x)\psi x \,.\, \sim(\exists x)\varphi x \,.\, \psi x : \supset : (\exists_7 x).\, \varphi x \vee \psi x$' is a logical truth, or a theorem in Russell's system. Of course, as it stands, we don't even need to prove the logical formula. We can easily tell that it is a theorem by looking at the indices of the numerical quantifiers and applying our basic arithmetical knowledge that $3 + 4 = 7$. So we use our basic arithmetic knowledge in order to verify the logical formula, and not vice versa.

But could we not prove the logical formula in Russell's calculus? To that end, we would have to write out the antecedent with sequences of 3 and 4 existential quantifiers with different variables, and then derive from it the consequent, written out with 7 existential quantifiers with different variables. (For deriving the consequent from the antecedent amounts to a proof of the conditional.) Thus, from a formula containing the variables *x*, *y*, *z* and *x*, *y*, *z*, *s* we could derive a formula containing the variables *x*, *y*, *z*, *s*, *t*, *u*, *v*. But how is that to be a proof of $3 + 4 = 7$? More poignantly, how should a formula with 700 different, but not numbered variables show me that $300 + 400 = 700$?

> It is not logic—I should like to say—that compels me to accept a proposition of the form (∃) (∃) ⊃ (∃), when there are a million variables in the first two pairs of brackets and two million in the third. I want to say: logic would not compel me to accept any proposition at all in this case. Something *else* compels me to accept such a proposition as in accord with logic.
>
> (*RFM* 155b; cf. 144–5)

First, if we allowed ourselves to *count* the variables on either side that would not be a calculation. Remember: the point of arithmetic (of addition, for example) is that you *don't* have to count the lot, but derive the result by calculation. This is the first part of Wittgenstein's objection: Arithmetic is not just a catalogue of true propositions, it is a method of calculation. Yet we can't calculate with logic (*LFM* 265, 285). Logic may allow us to construct and derive formulae *corresponding* to arithmetical equations, by containing certain numbers of variables on both sides of a conditional, but it doesn't tell what the numbers are. In a lecture, Wittgenstein gives the example of having to cater for the populations of Manchester and Liverpool. Suppose you know that Manchester has 530,000 inhabitants and Liverpool 467,000.

> Now what does Russell say about this? He says that in order to cater for both Manchester and Liverpool . . . you have to cater for all who are either inhabitants of Liverpool or inhabitants of Manchester.
>
> (*LFM* 274)

Thanks a bunch!

Secondly, even counting is of course to be reduced to logic alone. In the case of stroke notation we could try to resort to a sort of geometrical proof: by connecting every stroke on the left with a stroke on the right. But for one thing, nothing of the kind would be possible in a logical calculus, say of *Principia Mathematica*. And for another thing, it would be more in the nature of an experiment than a proof. As Wittgenstein explains:

> Now let us imagine the cardinal numbers explained as 1, 1 + 1, (1 + 1) + 1, ((1 + 1) + 1) + 1, and so on. You say that the definitions introducing the figures of the decimal system are a mere matter of convenience; the calculation 703000 × 40000101 could be done in that wearisome notation too. But is that true?—"Of course it's true! I can surely write down, construct, a calculation in that notation corresponding to the calculation in the decimal notation."—But how do I know that it corresponds to it? Well, because I have derived it from the other by a given method.—But now if I look at it again half an hour later, may it not have altered? For it is not surveyable [*übersehbar*].
>
> (*RFM* 144–5)

This is the core of Wittgenstein's objection to the logicist idea of reducing calculations to logical truths. A proof must be *surveyable* (*RFM* I §154: 95). By that Wittgenstein does not mean that it must be possible to take it all in at once, but merely that it must be easily reproducible and re-identifiable:[5]

> 'A mathematical proof must be perspicuous [*übersichtlich*].' Only a structure whose reproduction is an easy task is called a "proof".

5. Cf. Chapter 10.1. For a more detailed discussion of Wittgenstein's concept of surveyability, see Mühlhölzer 2006; Baker & Hacker 2009, 347–9; Marion 2011; for a somewhat different view, see Büttner 2016a.

> It must be possible to decide with certainty whether we really have the same proof twice over, or not. The proof must be a configuration whose exact reproduction can be certain.
>
> (*RFM* 143a)

For it to be possible to speak of proving (or calculating) that *X* equals *Y*, *X* and *Y* (as well as the intermediary steps) must be reliably re-identifiable. Thus in order to prove that *300 + 400 = 700* I must first of all have reliably applicable concepts of those numbers. In the stroke notation case, however, I'm just looking at some very long rows of strokes. I may perhaps, by drawing lines between two such sets, convince myself that they are equinumerous, but as I don't know the number of either of them, I shall not be able to apply this result to any other collections of objects I may encounter in future. The result is not transferable: I only know that this particular collection of strokes on this sheet of paper is equinumerous with that one. It's an experiment about physical objects, not a mathematical proof. A proof that *300 + 400 = 700*, by contrast, requires that I can easily reproduce the symbols in question, and know how to apply them.

Therefore Wittgenstein argues that abbreviations can be essential mathematical devices, not just trivial conventions of convenience (*RFM* 144ab). If in one system (e.g. stroke mathematics) it is impossible *reliably* to carry out, reproduce and check certain kinds of operations, the introduction of a new notation (e.g. positional number system) that allows such operations to be carried out and reproduced with ease and practical certainty is a substantive mathematical innovation: something substantially new that was not already contained in the primitive system. Similarly:

> if you had a system like that of Russell and produced systems like the differential calculus out of it by means of suitable definitions, you would be producing a new bit of mathematics.
>
> (*RFM* 176e)

And even if afterwards you can translate a proof in the new system back into a painfully long version of it in the original, unabbreviated system (with hundreds or thousands of strokes or variables), it is not the long version that does the work or provides a foundation. On the contrary:

> A shortened procedure tells me what *ought* to come out with the unshortened one. (Instead of the other way round.)
>
> (*RFM* 157b)

In short, as an attempt to give an epistemic foundation to arithmetic, the logicist reduction turns out to be circular. As Hao Wang put it:

> We are able to see that [a logicist reconstruction of an arithmetical equation] is a theorem of logic only because we are able to see that

> a corresponding arithmetic proposition is true, not the other way round.
>
> (Wang 1961, 335)

In Mathieu Marion's words:

> In order to understand the logical truth, one must introduce precisely arithmetical knowledge which is meant to be proved true by the logical truth.
>
> (Marion 1998, 230)

We use numerical quantifiers or count variables in order to identify a Russellian logical truth as such. If we then try to expand such a logical truth by expressing all arithmetical concepts, including numbers, by purely logical ones, the result will soon (even with two-digit numbers) be so unsurveyable that it ceases to be a proof of anything.[6]

3.6 Even If We Assumed (for Argument's Sake) That All Arithmetic Could Be Reproduced in Russell's Logical Calculus, That Would Not Make the Latter a Foundation of Arithmetic

Disregarding all the objections we have considered so far, assuming that *Principia Mathematica* provides an effective calculus whose theorems mirror precisely the results of arithmetical calculations,—why should it be thought epistemologically superior to ordinary arithmetic? Now we simply have two equivalent calculi: why should one of them be regarded as more fundamental (*LFM* 265)?—The idea is of course that Russell's calculus should be in some sense better because it's logic, and logic is thought to be 'the foundation of everything else' (*LFM* 266).

Wittgenstein objects to this view for two reasons. First, our logical derivations are not less error prone than our mathematical ones.

> We incline to the belief that *logical* proof has a peculiar, absolute cogency, deriving from the unconditional certainty in logic of the fundamental laws and the laws of inference. Whereas propositions proved in this way can after all not be more certain than is the correctness of the way those laws of inference are *applied*.
>
> (*RFM* 174e)

6. Sébastien Gandon suggests in Russell's defence that the definitions connecting the logical calculus with arithmetic can themselves be regarded as part of logic (Gandon 2012, 184), but it's not clear to me how that could amount to a rebuttal of the objection that what in Russell's calculus corresponds to an arithmetical equation fails to provide a proof of that equation (cf. Mathieu Marion's rejection of Mark Steiner's (1975) attempted defence of logicism (Marion 1998, 234–5)).

We may have an idealized image of the realm of logic as eternal laws 'of the purest crystal' (*PI* §97), but such a quasi-Platonic vision doesn't give our actual derivations in a logical calculus any super-human certainty. After all, every calculus sets standards of correctness by its procedural rules and definitions, and our judgements about the correctness of a given application of those rules is always fallible (especially when things get complicated), but can be practically indubitable (when we break it down to elementary steps)—independently of whether it is a calculus of quantifiers, connectives and predicates, or one of numbers and operations, or one about certain movements of wooden pieces on a board of 64 squares.

Secondly, there is no need to reduce mathematics to logic, because mathematics *is* logic, too.

> We can say that arithmetical propositions are laws of thought in the same sense in which logical propositions are.
>
> (*LFM* 267)

> In a perfectly good sense arithmetic is logic and also logic is arithmetic.
>
> (*LFM* 268)

That is to say, both formal logic and arithmetic are to give us rules for licit, truth-preserving transformations of linguistic expressions. Just as modus ponens allows us to move from two given propositions to a third, so an equation allows us to move from a number statement to another, e.g. from

'I have 12 sandwiches.'

to:

'I have as many sandwiches as it takes for 6 people to have 2 each.'

The difference between formal logic and arithmetic is merely that they provide rules for different groups of linguistic expressions: 'all', 'some', 'not', 'and', etc., in one case, and '1', '2', '3', '4', 'add', 'take away', 'distribute' etc. in the other. Neither formal logic nor arithmetic covers the whole of language. Formal logic has nothing to say on all the analytic relations between non-logical terms, and yet inferences based on analytic relations can be just as logical and compelling as those based on modus ponens.

So we can roughly distinguish three groups of logical inferences:

(i) those that are systematically codified in formal logic;
(ii) those that are systematically codified in arithmetic;
(iii) those that are not systematically codified.

Here is an example of group (i) with an abstract formulation of the relevant rule:

(24) Jones is in Cardiff or in Swansea. He's not in Cardiff. So he's in Swansea.
(25) $p \vee q, \sim p \vdash q$.

Similarly for group (ii):[7]

(26) I have 12 sandwiches. So I have as many sandwiches as it takes for 6 people to have 2 each.
(27) $12 \div 6 = 2$.

And an example of type (iii):

(28) Jones is a bachelor, so he is not married.

Now, logicism should not be described as an attempt to show that arithmetic is logical, i.e. a calculus or rule-governed system for transforming expressions. Of course it is. Primary school teachers explain how arithmetic equations follow from rudimentary definitions and practical explanations of numbers and arithmetical operators (cf. *PI* §28). Rather, logicism is an attempt to show that the rules of formal logic suffice to codify also the logical inferences based on the use of numbers and arithmetical terms. Would that in any way vindicate mathematical reasoning? No, our arithmetical reasoning is in no need of vindication. Its inferences and calculations are every bit as certain as those in formal logic (*BT* 584–5). The gain would merely be one of economy. Undoubtedly it affords a certain gratification, and structural insight, to be able to show how a complex system of rules can be derived from a very small number of basic rules. Thus it is interesting to see that all truth-functors can be defined in terms of a single one: the Sheffer stroke, or joint negation (as used by Wittgenstein in the *Tractatus*). But the re-definition of our familiar truth-functors of conjunction, disjunction, and conditional, in terms of joint negation alone doesn't make our calculus of propositional logic any more *certain* than it was before.

(Nor can it be said that only a maximally economical calculus of formal logic will give us a true understanding of the relevant logical inferences. Far from it. Axiomatic economy tends to be bought at the price of considerably greater complexity at the level of even the most humdrum inferences. Thus the clarity and self-evidence of even quite elementary logical inferences is lost when we express them with only a single truth-functor.)

7. Wittgenstein's views on the relation between mathematics and its application will be discussed in greater detail in Part II.

Likewise, a successful reduction of arithmetic to formal logic and set theory would of course be interesting in showing up links between two areas of logical reasoning, but it would in no way add to the certainty or reliability of arithmetic, which is a logical calculus in its own right, every bit as clear and trustworthy as that of formal logic.

4 The Development of Wittgenstein's Philosophy of Mathematics

Tractatus to *The Big Typescript*

4.1 *Tractatus Logico-Philosophicus*

Although the discussion of mathematics in the *Tractatus* is very laconic—a mere two pages—it contains the seeds of many key ideas of Wittgenstein's later writings. Already in his early writings, Wittgenstein regarded mathematics as logic: 'Mathematics is a method of logic' (*TLP* 6.234). Hence, as mentioned at the end of the previous section, he saw no need to *reduce* mathematics to logic (as Frege and Russell tried to do), since mathematics, as it stands, already *is* a kind of logic. And from an epistemological point of view, there is no need to translate one logical calculus into another. Repeatedly Wittgenstein characterises both logic and mathematics in the same way:

- Logical propositions, tautologies, say nothing (*TLP* 5.142, 5.43, 6.11); they have no sense, they express no thought; just as mathematical propositions do not express a thought (*TLP* 6.21).
- The logic of the world is reflected both by formal logic (in tautologies) and by mathematics (in equations) (*TLP* 6.22).
- In both logic and mathematics 'process and result are equivalent. (Hence no surprises.)' (*TLP* 6.1261; *NB* 24.4.15).
- In logic every proposition can be presented as its own proof (*TLP* 6.1265), 'one can recognize that [logical propositions] are true from the symbol alone' (*TLP* 6.113); similarly, every mathematical proposition must be self-evident (*TLP* 6.2341), their correctness can be ascertained without any comparison with the facts (*TLP* 6.2321).
- Logical constants don't denote logical objects (4.0312, 5.4); likewise number signs don't denote mathematical objects (*NB* 14.2.15).
- Logical propositions serve as inference rules for deriving empirical propositions from other empirical propositions (*TLP* 6.1201, 6.1264); and 'we make use of mathematical propositions only in inferences from propositions that do not belong to mathematics to others that likewise do not belong to mathematics' (*TLP* 6.211).

From that list we can draw as the three most fundamental characteristics of mathematical propositions that—like tautologies (cf. *NB* 19.10.14)—they are:

(a) non-descriptive,
(b) self-verifying (at least if presented transparently),
(c) rules for transforming empirical propositions.

These are the core ideas that provide the starting point for Wittgenstein's thoughts on mathematics when he returns to philosophy in 1929. Moreover, we shall see that the following tenets from the *Tractatus* doctrines continue to play a role in his later thoughts:

- Sense must be determinate (*TLP* 3.23).
- Every proposition has a complete analysis (*TLP* 3.25).
- Internal relations cannot be expressed in meaningful propositions (*TLP* 4.122).
- If a question can be asked, it can also be answered (*TLP* 6.5).

4.2 *Philosophical Remarks* (MSS 105–8: 1929–30) to *The Big Typescript* (TS 213: 1933)

The development of Wittgenstein's thoughts about mathematics in the manuscripts and typescripts to be published posthumously as *Philosophical Remarks* (1964) can be seen as motivated, and possibly triggered, by a realisation of the differences between tautologies and equations.[1] In particular, Wittgenstein came to retract claim (b) above, according to which equations are as self-verifying as tautologies.[2] In logic, proofs are not strictly necessary (*TLP* 6.1262); in a suitably transparent notation a tautology becomes recognizable as such. At least in the propositional calculus, proofs by deduction rules can be replaced by truth-value analysis (*TLP* 6.1203). Presumably Wittgenstein hoped that similar methods might be found for the predicate calculus. In mathematics, by contrast, line by line proofs (or calculations, e.g. to solve a quadratic equation) seem to be indispensable.

1. For a different account of Wittgenstein's key ideas on mathematics in his middle period, see Rodych 2018, §2.
2. It appears that, to begin with, Wittgenstein attempted to hold on to the idea that mathematical propositions are self-verifying by suggesting that 'the actual mathematical proposition is the proof' (*Der eigentliche mathematische Satz ist der Beweis*) (MS 105, 59), but of course that would just amount to an implausible redefinition of the term 'mathematical proposition'.

What is more, logical propositions should be altogether dispensable (*TLP* 6.211). And indeed, they play virtually no role outside logic classes. We have no need of the formula '$p \,.\, (p \supset q) \vdash q$' in order to apply modus ponens correctly, whereas we cannot effectively work out quantities of things (sizes or costs) without arithmetical calculations. This is an important asymmetry between logic and mathematics: the former is already implicit in our ordinary language, implicitly mastered by anybody who has mastered a language, the latter needs to be taught especially at school. The study of logical calculi (which of course can be more complicated than the logic implicit in everyday discourse) is of merely academic interest, but of no practical use in reasoning (hence not taught at school).

Finally, whereas one may be persuaded of Wittgenstein's view that the truth of a logical proposition can be recognized from the symbol alone, there are clearly mathematical propositions whose truth or falsity we are unable to ascertain (e.g.: 'Every even integer greater than 2 can be expressed as the sum of two primes'). There are mathematical conjectures, but there don't seem to be conjectures or unresolved problems in the predicate calculus (i.e. the logical resources invoked in the *Tractatus* to analyse ordinary language).

It was probably attention to those differences that changed Wittgenstein's view of mathematical propositions. In any case, he became more inclined to regard mathematical propositions as unlike tautologies and more akin to substantive propositions.

> It seems to me that you may compare mathematical equations only with significant propositions, not with tautologies.
>
> (*PR* 142)

But then a 'significant proposition' [*sinnvoller Satz*] has a sense, independently of whether it is true or false. It must be possible to understand it without yet knowing whether it is true or false. In the case of an empirical proposition one can understand what would be the case if it were true: one grasps the state of affairs described, which is what Wittgenstein regards as the 'sense' of the proposition (*TLP* 2.221). But of course a mathematical proposition is not a description of a state of affairs. So what could be its sense (or what could play a role similar to the sense of an empirical proposition)? Wittgenstein's answer was: 'the way in which it is to be proved' (*PR* 170; cf. *BT* 625). In short, if mathematical propositions have a sense and if proof is essential to mathematical propositions, it appears natural, from Wittgenstein's point of view, to identify the sense of a mathematical proposition with its method of proof.

In that way, by reflecting on crucial differences between logic and mathematics, Wittgenstein hit upon a particularly demanding kind of *verificationism* for mathematical propositions.[3] A meaningful mathematical proposition must have a proof (or disproof). We need not have written it down in order to claim an understanding of the proposition, but we must know the method of working it out. We must, in effect, be able to provide the proof (or disproof). Empiricist verificationism is less demanding in that respect. In order to claim understanding of an empirical proposition I must know what kind of evidence would verify or falsify it, but I need not actually be in a position to provide such evidence. The verificationist account of mathematics that emerged in *Philosophical Remarks* can be further characterised as follows.

3. It appears that it was only after Wittgenstein had developed a verificationist view in mathematics that he would also apply verificationist ideas to empirical propositions. The idea that only the proof of a mathematical proposition fully expresses its sense Wittgenstein first noted in MS 105, 59–60 (about February 1929); the first verificationist remark about non-mathematical propositions seems to occur in MS 105, 121 (written down after Wittgenstein had filled another volume, MS 106, and then returned to write on the verso pages of MS 105). The first general statement of verificationism explicitly applied to both empirical and mathematical propositions appears in MS 107, 143 (September or October 1929).—It has been suggested that verificationism was already implicit in the *Tractatus* (Wrigley 1989; but cf. Marconi 2000); it seems certainly true that the *Tractatus* could easily be developed to accommodate verificationist ideas (Hacker 1986, 138–41; Marconi 2002). However, that would apply only to empirical, not to mathematical propositions. *Pace* Wrigley (1989, 278–83), it seems very likely that Wittgenstein's new thinking about mathematics was influenced by L.E.J. Brouwer's lecture in Vienna on 10 March 1928 (Feigl 1981, 64; cf. Marion 2008). That Wittgenstein was to reject Intuitionism on the whole in no way contradicts the conjecture that *some* of Brouwer's ideas would have appealed to him, e.g. Brouwer's slogan: 'Mathematics is more an activity than a doctrine' (Becker 1964, 329; cf. *PR* 186e). An intuitionist inspiration for the mathematical verificationism of *Philosophical Remarks* is suggested by two pieces of textual evidence. Wittgenstein's most pithy formulation of mathematical verificationism is: '*Jeder Satz ist die Anweisung auf eine Verifikation*' (*PR* 174c). The published English translation ('Every proposition is the signpost for a verification') loses Wittgenstein's metaphor which is not of a signpost, but of a cheque or money order. And the first time Wittgenstein uses this metaphor for verificationism is in a remark commenting on one of Brouwer's claims (MS 106, 129). Moreover, this metaphor of a money order (*Anweisung*), of paper money (*Papiergeld*) as a mere promise of sterling currency was used in 1921 by another intuitionist mathematician, namely Hermann Weyl, whom Wittgenstein read and discussed repeatedly (cf. Marion 1998, 84–93). Weyl writes that a general mathematical statement is only a cheque (*Anweisung*) for particular judgements; and an existential theorem is merely paper money to be backed by a construction of a proof (Weyl 1921, 55). Only a few manuscript pages before the passage on Brouwer and verificationism, Wittgenstein explicitly rejects Weyl's claim about mathematical generality: 'The general equation is neither more nor less a judgement than the particular one (this is directed against Weyl)' (*Die allgemeine Gleichung ist nicht mehr und nicht weniger ein Urteil als die besondere [dies richtet sich gegen Weyl]*) (MS 106, 124). But he seems to have adopted Weyl's monetary metaphor in its wider application to theorem and proof.—For other aspects and further details of Brouwer's likely influence on Wittgenstein, see Marion 2008; for affinities between Brouwer's position and the *Tractatus*, see Marion 2003.

(i) From the *Tractatus* idea (c) that mathematical equations are transformation rules flows Wittgenstein's most fundamental characterisation of *mathematics as grammar*: that its equations are syntactical or, as he begins to call it in the 1930s, grammatical rules.

> Arithmetic is the grammar of numbers.
>
> (*PR* 130)

> An equation is a rule of syntax.
>
> (*PR* 143)

> The axioms—e.g.—of Euclidian geometry are the disguised rules of syntax.
>
> (*PR* 216)

> How are the equations of analysis connected with the results of spatial measurements? I believe, in such a way that they (the equations) fix what is to count as an accurate measurement and what as an error.
>
> (*PR* 217)

> geometry here is simply grammar.
>
> (*PR* 217)

(ii) If mathematical propositions are rules of grammar (either direct stipulations or what follows from them according to our derivation rules (*PR* 249)), their correctness or incorrectness cannot really be hidden from us or beyond our reach to ascertain. For the idea of a grammatical rule that nobody knows to be valid or would be able to ascertain as valid is nonsense:

> A law I'm unaware of isn't a law.
>
> (*PR* 176)

Therefore:

> we cannot have questions in mathematics that are in principle unanswerable. For if the rules of syntax cannot be grasped, they're of no use at all. And equally, it explains why an infinity that transcends our powers of comprehension cannot enter into these rules.
>
> (*PR* 143)

> I don't see how the signs, which we ourselves have made for expressing a certain thing, are supposed to create problems for us.
>
> (*PR* 185)

It's not only that questions in mathematics cannot be in principle unanswerable, they must be effectively answerable. We must know how to find out their answers; just as outside mathematics, cases of vagueness apart, it must always be possible to establish whether something is a grammatical rule of our language or not. Thus, Wittgenstein's seminal idea of mathematics as grammar also leads to *mathematical verificationism*. He writes with regard to mathematical propositions:

> Every proposition is the instruction/cheque [*Anweisung*] for a verification.
>
> (*PR* 174)[4]

> every significant proposition must teach us through its sense how we are to convince ourselves whether it is true or false. 'Every proposition says what is the case if it is true'. And with a mathematical proposition this 'what is the case' must refer to the way in which it is to be proved.
>
> (*PR* 170)

The sentence in inverted commas is an approximate quotation from the *Tractatus*:

> A proposition *shows* what is the case *if* it is true. And it *says that* it is the case.
>
> (*TLP* 4.022)

Note that the underlying *Tractatus* doctrine is not verificationism, but rather a form of truth-conditional semantics.[5] These are very different views, for, after all, in order to know what is the case if a proposition is true I don't have to know of ways of finding out. It is only when Wittgenstein's truth-conditional view is applied to the special case of mathematical propositions that a particularly demanding form of verificationism results.[6] For, as noted above, correct mathematical propositions have the peculiarity of being rules of grammar; but it seems absurd to suggest that we should not be able to say (after careful consideration) whether something is a rule of grammar or not. Furthermore, a grammar is at any given

4. The published translation ('signpost') is inaccurate (see n. 3 above).
5. Though different from more recent truth-conditional theories of meaning. See Hacker 1986, 324–9.
6. Of course he then extended his verificationism to all propositions, but in the empirical case it was less clearly motivated (cf. n. 3 above).

moment a closed system. We can of course extend it by introducing new rules of grammar, but we cannot *discover* new rules of grammar.

> The edifice of rules must be *complete*, if we are to work with a concept at all.—*We cannot make any discoveries in syntax*.—For, only the group of rules *defines* the sense of our signs, and any alteration (e.g. supplementation) of the rules means an alteration of the sense.
> (*PR* 182)

Hence:

> There are no gaps in mathematics. This contradicts the usual view. In mathematics there is no 'not yet'.
> (*PR* 187)

(iii) Moreover, since Wittgenstein had changed his mind about the status of mathematical propositions, regarding them no longer as only pseudo-propositions (*TLP* 6.2), but as something closer to meaningful propositions than to tautologies (*PR* 142), he was now inclined to say about mathematical propositions what in the *Tractatus* he said about propositions in general. Hence he would also apply to them his postulates of the *determinacy of sense* (*TLP* 3.23). In the *Tractatus* he demanded:

> A proposition must restrict reality to two alternatives: yes or no.
> (*TLP* 4.023)

Now he asks, in the same vein:

> Does a mathematical proposition tie something down to a Yes or No answer? (i.e. precisely a sense.)
> (*PR* 170)

Likewise, the *Tractatus* postulate of *complete analysability* is now applied to mathematical propositions:

> A proposition has one and only one complete analysis.
> [*TLP* 3.25]

But if we think of an ordinary arithmetical equation, such as '$25 \times 25 = 625$', it is clear that its complete analysis will *be* its verification. An incorrect equation can only seem right to us as long as we haven't really worked out the sum; as long as we haven't fully analysed it. But then, since a full understanding of a proposition's meaning would require us to know how to analyse it, a full understanding of a mathematical proposition's meaning would require us to be able to verify it: to give its analysis which is also its proof.

> *Every* legitimate mathematical proposition must put a ladder up against the problem it poses, in the way 12 × 13 = 137 does—which I can then climb if I choose.
>
> (*PR* 179)

> a mathematical proof is the analysis of the mathematical proposition.
>
> (*PR* 179)

Indeed, since the proof is just the fully analysed mathematical proposition, Wittgenstein says that the proof is the *actual* mathematical proposition, of which ordinary mathematical propositions are just abbreviations:

> The real mathematical proposition is the proof: that is to say, the thing which shows how matters stand.
>
> (MS 105, 59; quoted in editor's note at *PR* 184)

> The completely analysed mathematical proposition is its own proof. . . . a mathematical proposition is only the immediately visible surface of a whole body of proof and this surface is the boundary facing us.
>
> (*PR* 192)

Moreover, the completeness of mathematics follows not only from its status as grammar, but also from mathematical propositions' new status as meaningful propositions (propositions with a sense). For a sense cannot be incomplete (*TLP* 4.023):

> Mathematics cannot be incomplete; any more than a *sense* can be incomplete. Whatever I can understand, I must completely understand. This ties up with the fact that my language is in order just as it stands.
>
> (*PR* 188)

(iv) What we cannot straightforwardly work out to be true or false, has no sense. From this position it would appear to follow—implausibly—that there cannot be any serious mathematical problems. Wittgenstein acknowledges that that cannot be right:

> My explanation mustn't wipe out the existence of mathematical problems. That is to say, it isn't as if it were only certain that a mathematical proposition made sense when it (or its opposite) had been proved. (This would mean that its opposite would never have a sense

> (Weyl).) On the other hand, it could be that certain apparent problems lose their character as problems—the question as to Yes or No.
> (*PR* 170)

The first parenthesis seems problematic. Could a mathematical proposition have sense if its opposite, its negation, has been proved to be correct? Indeed, the second parenthesis seems to suggest that the answer must be No.[7]—The last sentence indicates that although Wittgenstein concedes that there must be some as yet unresolved mathematical problems, he is prepared to contradict the common view to some extent: He would be prepared to say that some things are wrongly regarded as mathematical problems. A case in point might be Fermat's proposition (*PR* 172).[8]

Immediately after the quoted remark Wittgenstein presents for the first time the rule-following problem (*PR* 171a). Then (*PR* 171b–d) Wittgenstein considers a solution to the problem of how mathematical problems might be possible: a preliminary proof of sense that does not prove the proposition to be true, but merely indicates a method of how such a proof could be constructed. However, the upshot of these three remarks is not entirely clear (cf. *PR* 180a). So, again:

> We come back to the question: In what sense can we *assert* a mathematical proposition?
> (*PR* 172)

Like a question, an assertion has sense only where the truth is not a foregone conclusion. That is, I must be able to understand the proposition in question without yet knowing it to be true:

> to be able to make an assertion, I must do so with reference to its sense, not its truth. As I've already said, it seems clear to me that I can assert a general proposition as much or as little as the equation $3 \times 3 = 9$ or $3 \times 3 = 11$.
> (*PR* 172)

What characterises such elementary equations is that we know how to work them out. Wittgenstein seems to suggest that such a systematic method of verification (or falsification) gives meaning to an equation to which it can be applied—independently of the result. Thus even a false

7. For clarification of Wittgenstein's reference to Weyl, see *WVC* 81–2.
8. 'Thus Fermat's proposition makes no *sense* until I can *search* for the solution to the equation in cardinal numbers.
   And "search" must always mean: search systematically.' (*PR* 175)

equation can be regarded as meaningful! Its meaning or sense is that it can be checked.

In general:

> Where there's no logical method for finding a solution, the question doesn't make sense either.
>
> Only where there's a method of solution is there a problem.
>
> (*PR* 172)

> We may only put a question in mathematics (or put a conjecture), where the answer runs: 'I must work it out'.
>
> (*PR* 175)

> And that, in general, is what a question is in mathematics: the holding in readiness of a general method.
>
> (*PR* 176)

(v) In order to justify his position, Wittgenstein also says that:

> 'search' must always mean: search systematically. Meandering about in infinite space on the look-out for a gold ring is no kind of search.
>
> (*PR* 175)

But that seems implausible. In ordinary parlance, a selective and unsystematic search would still be (describable and understandable as) a search. And, after all, Wittgenstein does not hold a correspondingly rigorous kind of verificationism for empirical propositions. 'There's a golden ring at the bottom of the North Sea' is meaningful, even though we are practically unable to verify it. And in this case, Wittgenstein would have no reason to deny that one could search unsystematically (by checking just a few places in the North Sea since one cannot check them all).

On the other hand, already in the *Tractatus* Wittgenstein was inclined to say, quite generally:

> If a question can be framed at all, it is also possible to answer it.
>
> (*TLP* 6.5)

This view is echoed in the following remark:

> What 'mathematical questions' share with genuine questions is simply that they can be answered.
>
> (*PR* 175)

(vi) Given his insistence on a straightforward method of answering a mathematical question, it is not surprising that Wittgenstein regards simple arithmetical equations as *paradigms* of mathematical propositions, or (where we haven't checked them yet) problems (see *PR* 172 quoted above). Other mathematical problems need to resemble them in being in the same way systematically solvable. But then it would appear that all mathematical problems must be fairly easy, as easy as 25 × 25, just 'pieces of homework' (*PR* 187b):

> Wouldn't all this lead to the paradox that there are no difficult problems in mathematics, since, if anything is difficult, it isn't a problem?
>
> But it isn't like that: The difficult mathematical problems are those for whose solution we don't yet possess a *written* system. The mathematician who is looking for a solution then has a system in some sort of psychic symbolism, in images, 'in his head', and endeavours to get it down on paper.
>
> (*PR* 176)

That is unconvincing.[9] Rather, the difficult mathematical problems are those with which the best mathematicians wrestle in vain for years or decades. And even if a problem is solved more quickly, it is hardly likely that the method of solving already existed in images in a mathematician's head. (Think, for example, of the complexity of Andrew Wiles's proof of Fermat's Last Theorem.)

(vii) It is not only that, on this account, what appears to be a mathematical proposition for which for the moment not even an unwritten method of solution can be produced does not qualify as a mathematical proposition. There is the further paradox that any proof to be discovered in future cannot count as a proof of that original proposition:

> If I hear a proposition of, say, number theory, but don't know how to prove it, then I don't understand the proposition either. This sounds extremely paradoxical. It means, that is to say, that I don't understand the proposition that there are infinitely many primes, unless I know its so-called proof: when I learn the proof, I learn something *completely new*, and not just the way leading to a goal with which I'm already familiar.
>
> But in that case it's unintelligible that I should admit, when I've got the proof, that it's a proof of precisely *this* proposition, or of the induction meant by this proposition.
>
> (*PR* 183)

Wittgenstein's response to this problem is based on a distinction between proper mathematics and mathematical 'prose' (*PR* 184a). The latter, our

9. Reminiscent of the way the author of the *Tractatus* at certain points gestures towards obscure mental processes of meaning in order to fill gaps in his philosophical picture.

informal talking *about* mathematics is not itself mathematics. Mathematics is a calculus; describing aspects of this calculus is not calculating, is not itself practising mathematics. Indeed:

> You can't write mathematics, you can only do it.
>
> (*PR* 186)[10]

And now Wittgenstein claims that the impression that there are as yet unsolvable mathematical problems arises only in 'prose', in our casual talk about mathematics, not in mathematics itself:

> Only in our verbal language . . . are there in mathematics 'as yet unsolved problems'.
>
> (*PR* 189)

Thus, he argues that in mathematics proper the question of whether there is a finite number of primes cannot be formulated—before a method of answering it has been devised (*PR* 188–9; cf. *PLP* 396).

> This is what creates the impression that previously there was a problem which is now solved. Verbal language seemed to permit this question both before and after, and so created the illusion that there had been a genuine problem which was succeeded by a genuine solution. Whereas in exact language people originally had nothing of which they could ask how many, and later an expression from which one could immediately read off its multiplicity.
>
> (*PR* 189)

Accordingly, in his discussion of Euler's proof of the infinity of primes (*BT* 645–9), Wittgenstein tried to offer an account of the prime numbers that would allow us to read off the concept's 'multiplicity' (cf. *TLP* 5.475).[11] He obviously continued to be attracted by the *Tractatus* idea

10. This is of course reminiscent of the *Tractatus* view that logic, internal relations generally, cannot be described in meaningful language. We can correctly infer that *q* from '*p*. (*p* ⊃ *q*)', but we cannot *say* that '*p*. (*p* ⊃ *q*)' entails '*q*'.
11. Wittgenstein's concern with Euler's proof is that it fails to give us a concept of a prime number (*BT* 648). Therefore, he proceeds to develop it so that it gives an upper bound for the next prime number: $p_{n+1} < 3p_n - 1$. This still falls short of the sort of expression we have, for example, for an even number, i.e. one which allows us to construct the series of even numbers and allows us to 'read off its multiplicity' (*PR* 189a). Still, the upper bound Wittgenstein provides turns the proof from a meaningless guarantee that we will find another prime number if only we search long enough, into a genuine proof that one of a finite number of numbers is prime. (I owe this observation to David Dolby.)—For a detailed discussion of Wittgenstein's reconstruction of Euler's proof, see Mancosu & Marion 2003; Lampert 2008.

that mathematical truth, like logical truths, should be made self-evident in a suitable notation. However, even if such transparency could be achieved in some cases, frequently we cannot do without lengthy calculations or proofs. And even in this case it would appear more plausible to say that we have a concept of a prime number and can raise questions about its extension *before* we find a proof that there are infinitely many prime numbers.

(viii) Of course the fact remains that people, even mathematicians, use such prose: they do form questions in verbal language before they are able to prove anything along those lines. How are we to account for that?

> How can there be conjectures in mathematics? Or better: what sort of thing is it that looks like a conjecture in mathematics? Such as making a conjecture about the distribution of primes.
>
> (*PR* 190)

Wittgenstein concedes that such conjectures, although not strictly speaking mathematical propositions or questions, have heuristic value:

> You might say that an hypothesis in mathematics has the value that it trains your thoughts on a particular object—I mean a particular region—and we might say 'we shall surely discover something interesting about these things'.
>
> (*PR* 190)

Such questions or conjectures give us new impulses, 'spurring us on to some mathematical activity', 'stimulating the mathematical imagination' (*Z* §§696–7; cf. *WVC* 144).

However, it doesn't seem plausible to say that the sense of a mathematical conjecture consists *merely* in its heuristic value: that 'it trains your thought on a particular region' of mathematics (*PR* 190). Later, considering Fermat's last theorem, Wittgenstein came to have doubts about this view (*RFM* 314c–e). After all, the concepts involved in this conjecture are all well-defined and used in what appears to be a straightforward and understandable way. So why should we not be able to attach a clear meaning to it? In some cases it could perhaps be argued that the conjecture combines those concepts in an unforeseen and as yet unintelligible way.

(ix) In 1933 Wittgenstein produced a book from the material of the ten manuscript volumes (MSS 105–14) he had written in the years 1929–32. This has become known as *The Big Typescript*. Beside numerous remarks already included in the 1930 collection posthumously published as *Philosophical Remarks* (which was based on the first four of those ten manuscript volumes) it contains further developments and refinements of the same ideas, among them his attempts at explaining mathematical research. Although his first tentative suggestion that the system in which a solution to an open problem can be calculated already exists in mathematicians' heads (*PR* 176) was clearly implausible, it would seem right to say that a non-trivial mathematical problem cannot be seen as a problem *in* mathematics. For in our existing mathematics, that is, our existing mathematical syntax or grammar, our system of calculations, such a problem has no solution. Hence a non-trivial mathematical problem must be seen as one of extending our syntax: of finding (or inventing) a system that preserves all existing techniques but also allows for a solution to the given problem (*WVC* 35f.).

What Wittgenstein seems to have in mind here is something like the question '4 – 6 = ?' before the introduction of negative numbers (or '7 : 3 = ?' before the introduction of fractions). In the existing system of natural numbers '4 – 6' has no solution and doesn't make sense. Hence the question 'How much is 4 – 6?' can only be understood as a request to find a suitable extension of our existing mathematical system by which this expression can be given a sense. So it's not a question *in* (our then existing) mathematics, but a 'prose' question about mathematics. Thus, Wittgenstein suggests, mathematical conjectures should be regarded as: 'signposts for mathematical research, stimuli for mathematical constructions' (*BT* 631; cf. *PR* 190).

In *The Big Typescript*, Wittgenstein retains the passage presenting the paradox that on his account there would appear to be no difficult mathematical problems; but then instead of his implausible initial suggestion of solutions pre-existing in mathematicians' heads (*PR* 176), he offers the following:

> What follows is, that the "difficult mathematical problems", i.e. the problems of mathematical research, aren't related to the problem "25 × 25 = ?" as, say, a feat of acrobatics is to a simple somersault (i.e. the relation isn't simply: very easy to very difficult). Rather, they are 'problems' in different senses of the word.
>
> "You say 'Where there is a question there is also a way to answer it', but in mathematics there are questions that we don't see any way to answer."—Quite right, but all that follows from that is that in this case we are using the word "question" in a different sense than in the case above.
>
> (*BT* 642; cf. *PLP* 397)

On this more tolerant view, problems of mathematical research are emphasised to be essentially different from homework problems, but nonetheless accepted as legitimate. So their formulation (as a conjecture, say) cannot be dismissed as nonsense. Wittgenstein illustrates the difference with an analogy:

> Imagine someone gave himself this problem. He is to invent a game; the game is to be played on a chessboard; each player is to have eight

> pieces; two of the white ones (the "consuls") that are at the ends of the beginning position of the game are to be given some special status by the rules; they are to have greater freedom of movement than the other pieces; one of the black pieces (the "general") is to have a special status; a white piece takes a black one (and vice versa) by being put in its place; the whole game is to have a certain analogy with the Punic wars. Those are the conditions that the game has to satisfy.—There is no doubt that that is a problem, a problem of a completely different kind from that of finding out how under certain conditions white can win in chess.—But now let's imagine the problem: "How can white win in 20 moves in the war game whose rules we don't yet know precisely?"—That problem would be quite analogous to the problems of mathematics (not to its problems of calculation [*Rechenaufgaben*]).
>
> (*BT* 620; cf. *PLP* 397–8)

In other words, a serious mathematical problem is like a problem in a game whose rules haven't been fixed yet! And the corresponding conjecture would be like the suggestion that in such a game in a certain position it would be possible for white to win in 20 moves.

Is that a plausible view? Probably not. The envisaged invention of a game involves free stipulations (about the permitted movements of different pieces); and it is clear that there will be numerous different solutions. By contrast, it would appear that the 'pieces' in, say, Fermat's Last Theorem are already sharply defined.

The mathematical situation Wittgenstein seems to have in mind here is one where a question leads us to define new mathematical symbols (or add to the rules for the use of existing symbols), comparable to new (or changed) pieces introduced in a board game ('consuls' and 'general'). For example, for someone who has learnt only calculations with natural numbers the question 'What is 5 – 7?' has no answer. For an answer to become possible, he first needs to introduce new symbols in his calculus: negative numbers.—However, mathematical problems are not always like that. For instance, the question whether there is a greatest prime number is (or was) a real mathematical problem, not just a routine calculation. And yet its solution doesn't require the definition of any new mathematical concepts or the introduction of any new mathematical techniques. It is much more like solving a problem in an existing game. After all, even the solution of a problem within a given framework of fixed rules may go beyond the simple application of existing algorithms and thus require a considerable amount of imagination.

More persuasive appears Wittgenstein's suggestion that when we seem to understand an unproven mathematical proposition we give it a corresponding empirical sense. For example, we can believe that however many numbers we try out, we will never come across any counterexample to

Goldbach's conjecture (*BT* 617–9). There are indeed examples of mathematical conjectures that although unproven are regarded as practically certain from an empirical point of view. Thus C.A. Rogers remarked that Kepler's sphere-packing conjecture was a claim that 'most mathematicians believe, and all physicists know' (Singh 1997, 314).[12]

(x) Already implicit in the *Tractatus* operation-based view of mathematics, but only spelt out later was Wittgenstein's emphatic anti-Platonism. Mathematics is not a description of a world of abstract objects, nor indeed of anything else (*TLP* 6.21). It is not descriptive at all:

> in mathematics the signs themselves *do* mathematics, they don't describe it.
>
> (*PR* 186)

This view is now applied with particular emphasis to the topic of infinity (not much considered in the *Tractatus*). On a descriptivist account, mathematical talk of infinity would be understood as a description of some infinite realm (e.g. an infinity of natural numbers regarded as abstract objects). Thus, infinity is construed as an extension. Wittgenstein, however, rejects the extensionalist idea of infinite totalities:

> you can't talk about *all* numbers, because there's no such thing as *all* numbers. . . .
>
> There's no such thing as 'all numbers', simply because there are infinitely many.
>
> (*PR* 147–8)

According to Wittgenstein, the sign for an extension is a list. Therefore, the notion of an infinite extension is incoherent. Infinity can only be understood intensionally: by a rule or general formula. The infinity, attributable to a certain intension, does not consist in the existence of a corresponding infinite extension, but in the unlimited possibility of generating corresponding (partial) extensions (*PR* 164).

Mathieu Marion observed that in his account of infinity Wittgenstein sided in a late-19th-century mathematical debate with Leopold Kronecker against Georg Cantor (Marion 1998, 1–10). Rejection of the extensionalist view of infinity amounts to a rejection of set theory with its use of infinite sets (*PR* 206). According to the extensionalist view, there is a set of all natural numbers, that is, the natural number series has a definite extension or size (*PR* 164), which means treating 'infinity'

12. As Timothy Gowers observes (2006, §6): 'A large part of mathematical research consists in spotting patterns, making conjectures, guessing general statements after examining a few specific instances, and so on. In other words, mathematicians practise induction in the scientific as well as mathematical sense.' Cf. Chapter 13.

as a number (or even as different numbers). Wittgenstein protests that to call the natural number series 'infinite' means, on the contrary, to deny that there is a definite number of natural numbers (*PR* 209). Rather, from any given number we can move on to its successor. Extensionalism, therefore, is guilty of a category mistake (Kienzler 1997, 147), confusing the absence of a number (or extension) with an unimaginably big number (or extension).

(xi) Another important topic that occupied Wittgenstein in the early 1930s was complete induction (or recursive proof). In line with his rejection of the extensionalist view of infinity, he rejects the idea that the argument of complete induction establishes a general claim about an infinite totality (*PR* 193, 201). The two steps $F(1)$ and $F(k) \supset F(k+1)$ do not establish that therefore all numbers are *F*; because there is no such thing as *all* numbers (*PR* 147). Rather, the two steps of an inductive argument are a recipe for showing how any particular number can be proven to be *F*: 'A recursive proof is only a general guide to an arbitrary special proof' (*PR* 196).

The main example Wittgenstein uses for his discussion of complete induction is an inductive proof of the associative law of addition given by Thoralf Skolem in 1923. The basis step is true by definition (*viz.*, the recursive definition of addition):

$$Df.: a + (b + 1) = (a + b) + 1. \qquad A(1)$$

Then assuming that what the definition states about 1, is also true for some number c (inductive hypothesis), Skolem proves in a few steps that it also holds for $c + 1$. Thus he has derived the inductive step:

$$\text{If: } a + (b + c) = (a + b) + c,$$
$$\text{then: } a + (b + (c + 1)) = (a + b) + (c + 1). \qquad A(c) \supset A(c + 1)$$

That then is to constitute a proof of the general algebraic rule

$$a + (b + c) = (a + b) + c. \qquad A(c)$$

i.e., the associative law of addition (*PR* 194–5).

In his early middle period, Wittgenstein is inclined to dispute that the proofs of $A(1)$ and $A(c) \supset A(c + 1)$ do indeed constitute a proof of the algebraic law $A(c)$.

For one thing, the algebraic formula $A(c)$ is not logically derived in the proof: it is 'not the end of the *chain of equations*' (*PR* 197). He rejects the view that the generality of the associative law is a new idea, something beyond the steps of the inductive proof, to which that proof directs us (*PR* 200). Reverting to *Tractatus* terminology, Wittgenstein says that the associative law, in its generality, is not a proposition (*PR* 198), not something that can be meaningfully stated, but it only *shows itself* in the proof (*PR* 203).

For another thing, Wittgenstein declares that the associative law of addition is a basic law of a new calculus, algebra, hence a stipulation and neither in need of nor susceptible of a proof (*PR* 193, 201–3; *BT* 724–5).

On the other hand, Wittgenstein acknowledges that there is some reason to accept the inductive steps as proof of *A(c)*.

> Yet [the recursive proof] surely does justify the application of A(c) to numbers. And so mustn't there, after all, be a legitimate transition from the proof schema to this expression?
>
> (*PR* 196)

It would appear that in some sense the inductive proof does establish a general conclusion. After all, the inductive proof serves me not merely as a recipe for producing new proofs, for example, a step-by-step proof for *A(7)*; 'it in fact spares me the trouble of proving each proposition of the form "A(7)"' (*PR* 196–7). *A(7)* can be derived directly from the algebraic formula *A(c)*, which therefore appears to function as a general statement after all. The following remark illustrates the tension in Wittgenstein's thinking at the time:

> An induction doesn't prove the algebraic proposition, since only an equation can prove an equation. But it justifies the setting up of algebraic equations from the standpoint of their application to arithmetic.
>
> (*PR* 201)

It doesn't prove it, but it justifies it. Isn't that a kind of proof?

Elsewhere Wittgenstein explains that the algebraic formula *A(c)* is best regarded as a sign denoting the inductive proof (*PR* 202). Or, again:

> We are not saying that when f(1) holds and when f(c + 1) follows from f(c), the proposition f(x) is *therefore* true of all cardinal numbers; but: "the proposition f(x) holds for all cardinal numbers" *means* "it holds for x = 1, and f(c + 1) follows from f(c)".
>
> (*BT* 675)

But that is not a reason to deny that the inductive argument *proves* the generalisation. After all, this is something that Wittgenstein holds about mathematical proof in general: that it determines the meaning of what it establishes (see (iii) above). In that sense then, it could be said that the laws of algebra are proven by arithmetic induction, not by formal logic perhaps (unless we include an inductive axiom in our logic), but as Wittgenstein suggests, in line with Jules Henri Poincaré (Waismann 1936, 86), as synthetic a priori truths (*BT* 672).[13]

13. For a discussion of other aspects of Wittgenstein's remarks on induction, see Marion & Okada 2018.

Another important topic is first touched upon in this context: the problem of rule-following: 'the unbridgeable gulf between rule and application' (*PR* 198). In order to understand the result of an inductive proof we have to see that a certain step can be iterated ad infinitum, while of course we cannot comprehend an infinity of steps in one insight. We have to trust ourselves always to be able to see what particular step will follow from a general rule. How can we be so sure? This aspect of inductive proofs Wittgenstein will return to in 1940 (MS 117). Accepting now that there are two types of proofs: mere calculations (*Ausrechnungen*), algorithmic derivations, on the one hand, and proofs that require some reflection, seeing some further implications in a derivation, on the other (MS 117, 165–6; cf. *BT* 710–21), Wittgenstein raises the challenge of the rule-following considerations: what if someone fails to see an inductive proof as justifying an instantiation at some later point? (MS 117, 168ff.)[14]—We shall come back to the rule-following problem in Chapter 7, and to Wittgenstein's later views on inductive proofs in Chapter 10.1(b).

To summarize, Wittgenstein's views on mathematics around 1930 are for the most part developments of ideas already present in the *Tractatus*. Like logic, mathematics is not descriptive, but a system of grammatical rules for the transformation of empirical statements. However, mathematics differs from logic in that it consists essentially of calculations and proofs. The sense of a mathematical proposition is its method of proof; that is Wittgenstein's mathematical verificationism, which, however, doesn't seem to provide a plausible account of mathematical research problems and conjectures. A corollary of the view of mathematics as normative, rather than descriptive is the rejection of Platonism. In particular, the concepts of mathematical infinities are not names of independently existing infinite totalities, actually infinite extensions as assumed in set theory. Mathematical infinity can only be understood as a rule for developing a series without an end point; inductive proofs should not be seen as deriving an extensional claim about *all* numbers of an infinite series.

There remains a tension in Wittgenstein's ideas in this period. He oscillates between *denying* that algebraic laws can be established by induction and *explaining* how they can be so established.[15] Later (perhaps from

14. For a detailed discussion of the MS 117 passages on induction, see Mühlhölzer 2010, 405–16.
15. There has been a scholarly disagreement between Stuart Shanker and Mathieu Marion as to whether Wittgenstein's discussion of Skolem's inductive proof is intended as an attack on Skolem (Shanker 1987, 199–210) or whether Wittgenstein is largely in sympathy with Skolem's approach (Marion 1998, 105–6: n. 36). I'm inclined to agree with Marion that Shanker overstates the polemical thrust of Wittgenstein's discussion and that there are considerable points of agreement with Skolem's ideas; on the other hand, Shanker is right in saying that, in some of his remarks, Wittgenstein denied that the associative law of addition could be *proven* by Skolem's inductive argument—although in other, more tolerant, remarks he appeared to allow it.

1934 on; see Waismann 1936, 89) he emphasises that algebra is a new calculus, not provable within arithmetic, without denying that it can be established by inductive proofs. In his later philosophy, far from seeing conceptual change as a reason not to speak of 'proof', he will emphasise conceptual change as a typical aspect of mathematical proof (see Chapter 10.1).

Part II

Wittgenstein's Mature Philosophy of Mathematics (1937–44)

5 The Two Strands in Wittgenstein's Later Philosophy of Mathematics

The canonical collection of texts presenting Wittgenstein's later philosophy of mathematics is *Remarks on the Foundations of Mathematics* edited by G.E.M. Anscombe, Rush Rhees, and G.H. von Wright, first published in 1956, and re-edited with additional material in 1978. It contains remarks written between 1937 and 1944 (when Wittgenstein stopped occupying himself with mathematics). Only Part I of that collection is a typescript edited by Wittgenstein himself (TS 222); all other parts are taken straight from manuscripts. The editors chose what sequences of those manuscript remarks they deemed most relevant and important, fully aware that some of their editorial decisions may have been contestable, occasionally leaving out remarks that should have been included. Still, on the whole it seems to me that the collection gives a fairly comprehensive picture of Wittgenstein's later philosophy of mathematics. Another important source are Wittgenstein's *Lectures on the Foundations of Mathematics*, held in Cambridge in 1939 and edited by Cora Diamond from the notes of four students. Although obviously less reliable in their details, those notes have the advantage that here Wittgenstein is trying to explain his ideas to others and sometimes responding to their reactions.

Wittgenstein's mature philosophy of mathematics hinges on two principal ideas, namely: that mathematics is essentially a calculus (or network of calculi) and that mathematics is essentially (akin to) a system of grammatical norms. In short: the *calculus view* and the *grammar view* of mathematics.

The intermediate period (roughly 1929–37) was dominated by the calculus view of mathematics. Here Wittgenstein put forward the claim that the meaning of a theorem is its proof. You can't understand a formula without understanding its position in the calculus, its derivation. Indeed, mathematics is primarily seen as the activity of calculating. Mathematical results are essentially the results of certain calculations.

As described in the previous chapter, such ideas led Wittgenstein to a demanding form of mathematical verificationism. To understand a mathematical proposition requires an understanding of how it can be calculated. What makes this more demanding than verificationism about empirical propositions is that whereas in the empirical realm you may be able to specify what evidence *would* verify (or falsify) a proposition, but find yourself contingently unable to lay your hands on such evidence, in mathematics, by contrast, if you know how something could be proved, you can prove it. That is because a mathematical proof can only be described once it has been discovered (*BT* 630–1). An appropriately detailed description of a proof is itself a proof.

Between 1937 and 1939, the emphasis in Wittgenstein's thinking shifted to the grammar view, which became the cardinal idea of his mature philosophy of mathematics. Therefore, the grammar view shall be explored in the following chapters (Chapters 6–8). However, Wittgenstein continued to hold on to the calculus view—at least to the core idea that mathematics essentially involves calculations and proofs—, and in Chapters 9 and 10 we shall turn to the difficulties he encountered in reconciling the two views.

The problem is this: Of course the acceptance of a norm of representation can be made conditional on its being derived according to certain procedural rules, but it is not clear in what sense such a derivation can be regarded as a proof. Comparing mathematical norms with laws, one may object that passing a law in a constitutionally correct manner is not proving it. If we take seriously the idea of proof (as showing to be *true*), the endorsement as a grammatical rule appears redundant. If a mathematical proposition is true, it is true independently of any such endorsement. If, on the other hand, such an endorsement is crucial (as the grammar view seems to suggest), should it not be possible even without a proof (cf. Kreisel 1958, 140)? Indeed, an unproven mathematical formula might find an application in physics. But then the calculus view would have to be qualified to allow for exceptions. On the grammar view, the meaning of a mathematical proposition would appear to be its use as a grammatical rule, whereas the calculus view seems to purport that the meaning of a mathematical proposition is given by its proof.

In order to resolve the tension, Wittgenstein suggests that the proof shows how a proposition can be applied (*RFM* 436e) and thereby justifies accepting it as a grammatical norm (which requires empirical applicability). It remains to be explored how plausible that is and how successful he was in combining the two strands of his thinking about mathematics: the grammar view and the calculus view.

6 Mathematics as Grammar

Are mathematical propositions descriptions of timeless abstract entities (Platonism) or are they generalisations of empirical observations (Mill; Formalism)? Neither, according to Wittgenstein. Platonism and empiricism share the assumption that mathematical propositions are *descriptions* of something, which is exactly what Wittgenstein rejects. They are not descriptions, they are *rules* (*RFM* 199a, 228f, 320b, 324b, *LFM* 33), or norms (*RFM* 425e, 431b): rules of expression (*LFM* 44, 47), representation (*RFM* 363c), or grammar (*RFM* 162d, 169b, 170b, 358be, 359a). They determine concepts (*RFM* 161c, 162b, 166bd, 172b, 248b, 320a, 432ef), or connexions between concepts (*RFM* 296cd, 432e), and hence the correct use of language (*RFM* 165h, 196f), a certain language game (*RFM* 236cd). Thus a mathematical proposition doesn't describe a fact (*RFM* 356ef), it merely determines what in a certain area of discourse makes sense and what doesn't (*RFM* 164bc).

There is a fairly uncontroversial sense in which some mathematical propositions can be called 'rules'. First, there are the most elementary arithmetical sums. '*1 + 1 = 2*', '*2 + 1 = 3*', '*3 + 1 = 4*', etc. serve as something like definitions, or basic paradigms of the use of natural numbers and the operation of addition: basic rules of the arithmetical calculus. Secondly, there are non-basic, but simple equations, such as in the times tables, which we memorize at an early age and apply when doing longer calculations.[1] Thus, '*3 × 9 = 27*' is applied as a rule when working out the long multiplication: *399 × 39* (cf. *PLP* 53). Thirdly, at a slightly more advanced level there are proven algebraic formulae that are memorised or consulted for repeated application, e.g. the cosine rule or the quadratic formula. However, such cases are *not* what Wittgenstein has in mind when he calls mathematical propositions rules. 'If one says the mathematical proposition is a rule', he writes, 'then of course not a

1. In his 1931–32 lectures Wittgenstein called them 'definitions' (*LL* 96).

rule in mathematics' (MS 127, 236).[2] Rather, on his view, they are rules of grammar, and, what is more, not the grammar of mathematics, but the grammar of non-mathematical language. As explained in Chapter 4.1, the idea goes back all the way to the *Tractatus*, where mathematics is characterised by the following three propositions:

> Mathematical propositions are equations.
>
> (*TLP* 6.2)

> If two expressions are combined by the sign of equality, that means that they can be substituted for one another.
>
> (*TLP* 6.23)

> in real life . . . we make use of mathematical propositions only in inferences from propositions that do not belong to mathematics to others that likewise do not belong to mathematics.
>
> (*TLP* 6.211)

Thus, the equation '2 + 3 = 5' is a grammatical rule for the use of number words in a natural language, licensing, for instance, the inference from 'I have two coins in my right pocket and three coins in my left pocket' to 'I have (at least) five coins in my pockets'.

Wittgenstein lays particular stress on the dependence of mathematics on its having applications *outside* mathematics. That is what turns a mere calculus, a game of manipulating signs according to certain syntactic rules, into mathematics.

> it is mathematics, I should think, when it is used for the transition from one proposition to another.
>
> (*BT* 533)

> It must be essential to mathematics that it can be applied.
>
> (*BT* 566)

> I want to say: it is essential to mathematics that its signs are also employed in *mufti*.
>
> It is the use outside mathematics that makes the sign-game into mathematics.
>
> (*RFM* 257de; cf. 295f)

2. *Wenn man sagt, der mathematische Satz ist eine Regel, so natürlich nicht eine Regel in der Mathematik.* (MS 127, 236; post 4.3.44)

> mathematical propositions containing a certain symbol are rules for the use of that symbol, and . . . these symbols can then be used in non-mathematical statements.
>
> (*LFM* 33; cf. 256)

Equations, according to Wittgenstein, are rules for connecting concepts, thus forging a new enriched concept (*RFM* 412f, 432f). The grammatical rule this provides is that where one side of the equation applies, the other one must apply, too: If, for instance, something is *2* + *3*, it must also be *5*. In this way, the equation '*2* + *3* = *5*' provides a further determination of the concept '*2* + *3*' (*RFM* 320a).

The normative status of mathematical propositions explains their peculiar necessity. It is the necessity of conceptual stipulation and its implications, which we find so much more inexorable than the necessity of any laws of nature. The latter we can establish only inductively, which means that it always remains at least conceivable that at some point the observed regularity will break down. Conceptual necessity, however, is not dependent on the way things are and continue to be in the world, it is fixed by ourselves: by the meanings we give our signs and on which we insist. Bachelors may at some point be able to breathe without lungs, or even cease to obey the laws of gravity, but they cannot fail to be unmarried men—simply because we are determined not to apply the predicate 'bachelor' to somebody who isn't an unmarried man. Should we at some point decide to do so then the word will have changed its meaning, it will *eo ipso* express a different concept, and the analytic claim that a bachelor is an unmarried man does of course presuppose that the words are used with their current meaning. Similarly, in mathematics the word 'must' conveys that we insist on a given concept (*RFM* 238d, 309j), independently of what experience may teach us (*RFM* 239d).

> The mathematical Must is only another expression of the fact that mathematics forms concepts.
>
> (*RFM* 430f)

> The emphasis of the *must* corresponds only to the inexorableness of [our] attitude both to the technique of calculating and to a host of related techniques.
>
> (*RFM* 430e)

However, the claim that mathematical propositions are rules of grammar provokes the question, first asked by Georg Kreisel (in his notoriously unsympathetic review of Wittgenstein's *Remarks on the Foundations of Mathematics*): why in that case proofs should be needed, 'since a rule of language, as ordinarily understood, is a matter of simple decision'

(Kreisel 1958, 140).—Well, this is a bit like asking: 'If tennis is a ball game, then why can't one score any goals?' The answer is: It's not analytic that ball games involve goals. It may be that the most popular ball games do; but there are others that don't. Similarly, it's not analytic that a grammatical rule must be arbitrarily chosen. Perhaps most of them are; but, on Wittgenstein's account, there is at least one kind of grammatical rule (in what we call mathematics) that is not arbitrarily chosen, but introduced by proof.

It is of course true that elsewhere Wittgenstein calls grammar 'arbitrary' (*PG* 184; cf. *PI* §372). But what he means by that is that it is essentially a human artefact that cannot be assessed as true or false to nature. That, however, does not mean that anybody will be allowed to introduce any kind of grammatical rule as the whim takes him. There may well be certain grammatical rules which—although we make them—we make only according to certain rule-governed procedures, which we call proofs.

Indeed, the idea of a practice with second-order rules that are not stipulated from the beginning, but can be, or have to be, introduced later according to certain rule-governed procedures is not at all unheard of. A common example is a legal system, which consists not only of a set of laws, but also of procedural rules for legally introducing new laws.

So, according to Wittgenstein, mathematical propositions have two key features: (i) They are given the status of grammatical rules, norms of representation. (ii) Definitions (and axioms) apart, they have to be legitimised by proof. The second feature will be considered in a later chapter (Chapter 10); now we are going take a closer look at the idea that mathematics is a kind of grammar.

As grammatical statements, mathematical propositions are said to provide a framework for descriptions, not to describe anything themselves. They determine what makes sense, but do not establish any substantive truth:

> For the mathematical proposition is to show us what it makes SENSE to say.
>
> (*RFM* 164b)

> If you know a mathematical proposition, that's not to say you yet know *anything*. I.e., the mathematical proposition is only supposed to supply a framework for a description.
>
> (*RFM* 356f)

This is provocative, and Wittgenstein himself at times felt provoked by it. He acknowledged that there appears to be a tension between this claim—what might be called a non-cognitivist account of mathematics—and mathematics' well-known prognostic potential and practical

usefulness. Equations are to be mere transformations of expressions, but 'How can the mere transformation of an expression be of practical consequence?' (*RFM* 357a).

> I can use the proposition '12 inches = 1 foot' to make a prediction; namely that twelve inch-long pieces of wood laid end to end will turn out to be of the same length as one piece measured in a different way. Thus the point of that rule is, e.g., that it can be used to make certain predictions. Does it lose the character of a *rule* on that account?
>
> (*RFM* 356a; cf. 381a)[3]

Very often I can calculate what will happen: mathematics teaches me an observable result. An area of 7 by 5 foot is to be covered with tiles each 1 foot square. How many tiles are required? An elementary calculation tells me that I shall need 35 tiles. So it appears that mathematics can serve to discover an empirical truth, and not just to fix sense.

Moreover, if an equation were just a rule to determine what makes sense and what doesn't, the application of a miscalculation should result in nonsense. For example:

> (A1) In order to cover an area of 7 by 5 foot I need to fit 37 tiles of 1 foot square.

—should make no sense; but it seems much more natural to say that it's just false. After all, I can wonder if it might not be true (*RFM* I §67: 62); I can try and convince myself empirically that I don't need 37 tiles. And what's more, it is not inconceivable that (A1) *might* be true. We could perhaps imagine that somehow when we put down 35 tiles there still remain two empty squares; and, strangely, when we count the laid out tiles we always get 37 (cf. *RFM* I §137: 91b).

In the early 1930s, Wittgenstein (as reported by Waismann) denied that a mathematical proposition allows us to predict empirical observations:

> It may seem as if the equation 5 + 7 = 12 entitles us to make statements about the future, namely to predict what number of shillings I shall find if I count the ones I have in each pocket [5 and 7]. But this is not the case. Such a statement about the future is justified by a physical hypothesis which stands outside the calculus. If a shilling suddenly disappeared, or if a new one suddenly came into existence while we were counting, we should not say that experience had

3. Cf. MS 163, 62r: '"Mathematics grammar?? But it helps us to make predictions!"—It *helps* us'. ("*Die Mathematik eine Grammatik?? Aber sie hilft uns doch Vorhersagen machen!"—Sie hilft uns.*)

> disproved the equation 5 + 7 = 12; similarly, we should not say that experience had confirmed the equation.
>
> (*PLP* 51–2)

On this view, mathematics only connects the concepts of *5 + 7* and of *12*, forming and insisting on a joint concept, according to which whatever falls under the one falls under the other. Using the joint concept to move from one description to the other does not involve any physical hypothesis. If you have 5 and 7 shillings in your pockets, then *ipso facto* you have 12 shillings in your pockets. But things are different if we talk about predicting what *will* be found in your pockets the next moment. The mathematical substitution rule '*5 + 7 = 12*' alone cannot guarantee that having counted 5 and 7 shillings *now*, I shall count 12 shillings a moment later. That prediction also involves the physical hypothesis, or assumption, that coins have a certain durability: that they don't suddenly disappear, coalesce or multiply.

One might reply, however, that although the durability of coins must of course be presupposed, the fact remains that the prediction in question is arrived at and justified (at least partly) by the equation. When asked: 'What makes you think that you'll find 12 shillings in your pockets', the answer is not (or not merely): 'What I know about the physical nature of coins', but rather: 'I counted 5 in one pocket and 7 in the other, and *5 + 7 = 12*'.

The question remains: How can Wittgenstein's view of mathematical propositions as rules of grammar be reconciled with their prognostic usefulness? But are they really grammatical propositions (cf. Marion 1998, 4)? If we compare elementary mathematical propositions with ordinary grammatical propositions—such as:

> (GP) A bachelor is an unmarried man.

—it soon becomes clear that they are significantly different. (GP) is constitutive of the meaning of its subject term: it explains what the word 'bachelor' means. 'Bachelor' and 'unmarried man' are just two expressions for the same concept. Hence, if you understand the expressions, you cannot ever know that one of them applies without knowing that the other one applies as well. By contrast, (as famously pointed out by Kant)[4] *7 + 5* and *12* are different concepts: they have different criteria of application (counting to 7 and counting to 5 versus counting to 12)

4. Kant 1787, B 15: 'But if we look more closely we find that the concept of the sum of 7 and 5 contains nothing save the union of the two numbers into one, and in this no thought is being taken as to what that single number may be which combines both. The concept of 12 is by no means already thought in merely thinking this union of 7 and 5'.

(cf. *RFM* 357f). Hence it is *possible* to find on a given occasion that one criterion is fulfilled, while the other one is not: to count 7 and 5 objects, but then to count only 11 altogether (or, to use Wittgenstein's example, *25 × 25*, but not *625*) (*RFM* 358ef). In this case we have, initially, two distinct concepts, independently comprehensible—'Only through our arithmetic do they *become one*' (*RFM* 358b; cf. 359a). Note the emphasis on 'become': If mathematical propositions are grammatical propositions they are essentially *additional* ones: *further* rules for terms that are already understandable without them. Mathematical propositions *enrich existing meanings*. The norm expressed by a grammatical proposition like (GP), by contrast, does not *change* or *enrich* the meaning of the word 'bachelor', it gives it its meaning in the first place.

It would appear that mathematical propositions are more like another type of grammatical proposition, fairly common in scientific discourse. Sometimes what used to be an empirical discovery is later made part of a definition, for example, the velocity of light or the key properties of an acid. Thus, Wittgenstein writes that arithmetical propositions originate from empirical observations that are at some point turned into rules: an arithmetical proposition is so to speak 'an empirical proposition hardened into a rule' (*RFM* 325b). In general:

> Every empirical proposition may serve as a rule if it is fixed, like a machine part, made immovable, so that now the whole representation turns around it and it becomes part of the coordinate system, independent of facts.
>
> (*RFM* 437e)

On this view, a mathematical proposition has been grafted onto a corresponding empirical observation.[5] By contrast, it could never have been empirically discovered that a bachelor is an unmarried man.

If elementary mathematical propositions are essentially additional rules for combining existing concepts, the question is whether these rules become fully integrated in our language, as Wittgenstein seems to suggest when he calls them 'grammatical' or 'instruments of language' (*RFM* 162d, 164–6, 358d, 359a), and when he says that mathematics 'moves along the rules of our language' (*RFM* I §165: 99), and 'forms ever new rules' (*RFM* I §166: 99), and that mathematics is 'deposited among the earliest [*Ur-*] measures' (*RFM* I §165: 99). There are, I believe, reasons to return a more qualified answer: reasons not to regard mathematics—except perhaps for its very rudiments—as part of the grammar of our *ordinary* language.

5. This idea will be critically discussed in Chapter 9.

What characterizes a grammatical proposition is that, as it determines what makes sense, its negation, or a sentence that violates the norm it expresses, is nonsense. Is that also true of mathematical propositions? As quoted earlier, Wittgenstein seemed to think so (*RFM* 164b). And at the most elementary level this may indeed be so. The sentence 'I had two coffees in the morning and two in the afternoon, so I had only three overall today' is patently inconsistent. It might well be dismissed as not only false, but nonsensical. But suppose someone said:

> (A2) The pitch of the roof of my lean-to garage is 15° to the horizontal and the roof extends 5.36 metre horizontally from the wall, and one side of the roof is 1.32 metre higher than the other.

Would we dismiss *that* as nonsense? Certainly not straightaway, for as far as we know it might even be true. Only a trigonometric calculation shows that:

> (M2) If a right-angled triangle has an angle of 15° and the adjacent side is 5.36 metre then the opposite is 1.44 metre.

So (A2) cannot be correct after all. And yet one can *believe* it to be correct—which speaks against regarding it as nonsense. For where there is no sense, there is nothing to believe. And yet, one can hold inconsistent beliefs. What, in such a case, does one believe if the sentence expressing one's beliefs doesn't make sense?

Wittgenstein discusses the issue of believing false equations (*RFM* 76–9), and seems to suggest that when I mistakenly believe that *16 × 16 = 169*, I do not believe a mathematical proposition; rather, I mistakenly believe that '*16 × 16 = 169*' *is* a mathematical proposition, a rule of our mathematical grammar—which it isn't. It has no meaning *in arithmetic*, just as moving a pawn backwards is not a move in chess, not even a bad one.

However, our question was not how one could believe a false *mathematical* proposition, but how one could believe a false *non-mathematical* proposition such as (A2) that is in conflict with a mathematical norm of representation (M2) and should be ruled out by it. Of course, it is not impossible for a trigonometric norm of representation to be put in terms of the shape of the roof of a lean-to garage, but that is hardly a natural understanding of (A2). The corresponding trigonometric norm would more naturally be put in the form of a conditional, like (M2). It is much more likely that we take (A2) as an empirical statement: as the speaker's report of his measurements. I can certainly believe that those are the correct measurements *before* I've done the maths. Afterwards I shall think that the speaker must have made a mistake (or that the roof isn't straight, so that one cannot really speak of a pitch of 15°). Yet the fact remains that one can understand (A2) as an empirical statement and believe it to be true.

Perhaps the case is analogous to that of believing that $16 \times 16 = 169$. There I believe that something is a mathematical proposition, although it is nonsense; here I believe that something is an empirical proposition, i.e. something that could be true, although it is in fact inconsistent. Such a situation may also arise where no mathematical concepts are involved. For instance, family relation concepts may be combined in an inconsistent manner: 'My son did not realise that neither of my grandniece's parents have any cousins'. When we believe such an account to be true we may wrongly believe that the description is consistent.

So, believability is not a guarantee of empirical sense. However, Wittgenstein himself seems to suggest that the applications of mathematical calculations may or may not be in agreement with experience, that is, that they are, or can be taken as, empirical claims.

In his 1939 lectures Wittgenstein points out emphatically and repeatedly that corresponding to an arithmetical equation there is an empirical statement expressed in similar or even the same words that it is important to distinguish from the mathematical proposition (*LFM* 111). One might be inclined here to think of a pair like:

(M) $5 + 7 = 12$
(A) 5 apples and 7 apples are 12 apples.

But as so often in his philosophy, Wittgenstein tells us not to look only at forms of words, but at the use made of them (cf. *LC* 2). The occurrence of an empirical term like 'apple' is no reliable indication that we are considering an empirical statement. As he remarks elsewhere, 'mathematical propositions might quite well be expressed in terms of people, houses, or what not' (*LFM* 116; cf. 113). A norm of representation can be taught by giving a possible instantiation. A term such as 'apple' may function somewhat like a variable, indicating that an arithmetic equation is essentially applicable to things, and not, as Platonists have it, a self-sufficient statement about abstract objects.

So, although apparently about apples, (A) can well be used as a mathematical proposition: as an expression of a norm of representation. On the other hand, the naked equation (M) could, according to Wittgenstein, be taken as an empirical generalisation. So the distinction between the mathematical and the non-mathematical use of number sentences need not coincide with that between the two kinds of formulations, but can cut right across it. Whatever formulation you choose it can be understood either way:

> The point is that the proposition "$25 \times 25 = 625$" may be true in two senses. If I calculate a weight with it, I can use it in two different ways.
>
> First, when used as a prediction of what something will weigh—in this case it may be true or false, and is an experiential proposition.

> I will call it wrong if the object in question is not found to weigh 625 grams when put in the balance.
>
> In another sense, the proposition is correct if calculation shows this—if it can be proved—if multiplication of 25 by 25 gives 625 according to certain rules.
>
> It may be correct in one way and incorrect in the other, and vice versa.
>
> It is of course in the second way that we ordinarily use the statement that 25 × 25 = 625. We make its correctness or incorrectness independent of experience. In one sense it is independent of experience, in one sense not.
>
> (*LFM* 41; cf. 292)

The point to note is that, on Wittgenstein's account, for all mathematical propositions (at least at the level of everyday mathematics) there are analogous empirical statements to the effect that things will, as a matter of fact, turn out in accordance with the mathematical norm (*LFM* 111). For (M) '*5 + 7 = 12*', for instance, there is the prediction that if you count 5 apples and 7 apples then counting the total will indeed yield 12. That prediction may occasionally, rarely, be false.

Of course, the applied statement (A) could also be taken in a timeless sense, in which it would be as unfalsifiable as (M). Whether after having added 7 apples to 5 apples the next moment there will be 12 apples is indeed an empirical matter (cf. *RFM* I §37: 51), but that the apples in question, being 7 and 5, are 12 in total is not. That is exactly Wittgenstein's point mentioned above that even a statement about apples can be taken, timelessly, as a mathematical one (*LFM* 113). But then it isn't really an empirical application yet. Where we apply maths to the empirical world we make claims about what the result of certain counts or measurements would be, yet the actual result of counting or measuring, at a certain point in time, is always an empirical matter. In an example of the kind alluded to in the passage quoted above, it is not inconceivable that 25 nuts weighing 25 grams each turn out to weigh 623 grams in total. Likewise, it remains conceivable that our measurements confirm (A2)[6]—in spite of its incompatibility with (M2).[7]

It appears then as if the same sentence could be used to express what must be the case and also what only happens to be the case. How is that possible? To return to the more primitive example (A), we can regard it

6. (A2) The pitch of the roof of my lean-to garage is 15° to the horizontal and the roof extends 5.36 metre horizontally from the wall, and one side of the roof is 1.32 metre higher than the other.
7. (M2) If a right-angled triangle has an angle of 15° and the adjacent side is 5.36 metre then the opposite is 1.44 metre.

as a necessary truth that 5 apples added to 7 apples amount to 12 apples or we can convince ourselves empirically that that is so, thereby allowing for the possibility of another result. By contrast, in the case of an ordinary grammatical proposition (such as [GP] ‘A bachelor is an unmarried man’) there is no room for any empirical confirmation. Having identified somebody as a bachelor there can no longer be a question as to whether he is an unmarried man since the two expressions are identical in meaning: have exactly the same criteria of application (unlike ‘*5* + *7*’ and ‘*12*’).

The difference springs from the fact, already observed, that arithmetic provides *additional* grammatical rules. In other words, there is already meaningful language with criteria of correctness *before* the advent of arithmetical rules. More specifically, number words have meaning, determined by the practice of transitive counting—before we develop or learn the rules of arithmetic: the techniques of addition, subtraction, multiplication, and division. At that stage we can count 5 apples and count 7 apples; and we can also count 12 apples. But whether where we counted first 5 and then 7, we will count 12 in total is still an open question, for the two concepts are different (cf. Baker & Hacker 2009, 322–3). Only in the new system of arithmetic the concepts of *5* + *7* and of *12* become identified (*RFM* 358b). In other words, we have here an overlap of what can be regarded as two (or more) different languages of increasing richness. First, there is what can be called ‘$_{N}$English’, that is to say: *ordinary English, including number words*, but no arithmetical operations; then, at primary school we learn elementary arithmetic. The result can be described by saying that now we have mastered ‘$_{N+}$English’ (i.e., *English with arithmetic*). Arithmetical equations are grammatical propositions in $_{N+}$English (just as in $_{N+}$French or $_{N+}$German, of course), their applications to numerical statements about apples are necessary truths, whereas other numerical statements, in conflict with them, are ruled out as nonsense. However, from the more primitive perspective of $_{N}$English, both such applications and their contraries (e.g. ‘When you add 5 apples to 7 apples, you have 11 apples’) are merely empirical claims.

Subsequently, further mathematical suburbs are added to our language (cf. *PI* §18). Historically, trigonometry begins to be developed in late antiquity; ontogenetically, it is at secondary school that trigonometric concepts are added to our linguistic repertoire. That can be described by saying that we learn to master $_{N+\Delta}$English: i.e., *English with arithmetic and trigonometry*. However, very few people continue to use, and hence to remember the rules of, goniometric functions in their adult life. Most people would not be able without further instruction to calculate (M2) and thus to determine the inconsistency of (A2) in $_{N+\Delta}$English. So, naturally, they would be inclined to treat (A2) as an empirical statement.

Thus, there are three differences between mathematical propositions and ordinary grammatical propositions. First, since in arithmetic we calculate what can also be counted, we have different criteria for identifying the same quantity (e.g. counting first 7 and then 5—or counting 12; or again, counting 3 columns of 8 rows—or counting 24 items), that is, different procedures of verification of a number statement, which a calculation shows to be equivalent. Yet different procedures of verification can always be imagined to yield different results, at least due to inaccuracy:

> there is no contradiction in saying: "By one method of counting I get 25 × 25 (and so 625), by the other not 625 (and so not 25 × 25)". Arithmetic has no objection to this.
>
> (*RFM* 358f)

By contrast, an ordinary grammatical proposition, spelling out the implications of a given concept, doesn't go beyond the criteria of application of the concept in question and therefore leaves no room for conflicting empirical results.

Secondly, most equations must be *calculated* should the need arise and are not memorised once and for all. Therefore, often even a competent mathematician cannot see immediately whether a quantitative claim is legitimate or incongruous. That produces temporarily something akin to an empirical attitude towards a proposition that ultimately turns out to be either necessary or inconsistent. Provisionally we regard as an empirical claim what later we may reject as nonsense.

Occasionally such a situation may also arise outside mathematics. A logically complex set of propositions (perhaps premises and conclusion of a fallacious argument) may involve a hidden inconsistency. But then, such a complex and non-obviously inconsistent expression, although it may be thought to be an analytic truth (e.g. if it's the conditional formed from premises and conclusion of an argument), it would not be regarded as a grammatical proposition. For what Wittgenstein calls a grammatical proposition, is characterised by its normative function: as a sample, reminder, or standard of conceptual correctness (*AL* 31). Yet if a proposition or set of propositions is so complicated that we cannot easily ascertain and need to work out whether it is analytic or inconsistent it would thereby be quite unsuitable to be invoked as a standard of correctness, or as a tool for explaining concepts—as a grammatical proposition such as 'A vixen is a female fox'.

Finally, there is a third difference between mathematical propositions and ordinary grammatical propositions. With the exception of basic arithmetic, mathematics is a specialist skill; hence most of its rules and results do not govern our vernacular, but merely a sub-language used by a small number of specialists (cf. RR 128). Therefore, even when ruled out by a mathematical norm of representation a quantitative statement, such as (A2), can still function as an empirical claim in ordinary language. That is to say, not only could (A2) be the result of our measurements, but we might accept it as correct; whereas a competent trigonometrist might also note down the measurements given in (A2), but would ultimately conclude that something had gone wrong (be it that he had measured inaccurately, or that the tape measure was faulty, or that some of the boards weren't straight or had shifted). Similarly, we may indeed find that after putting down first 7 and then 5 apples there are only 11 on the table; but then we would insist that 12 had been put there and one must somehow have disappeared (cf. *RFM* I §37: 51).

Although mathematical propositions are somewhat different from ordinary grammatical rules, we can still agree with Wittgenstein that they function as norms of representation. They define a standard of correctness by which sentences can be assessed. Thus,

(N) I bought 25 bags of 25 nuts each; I had 623 nuts in total.

—can be rejected, not on empirical grounds, but a priori; just as we know a priori that 'Jones was a married bachelor' is unacceptable. The difference is only that the latter is nonsense, whereas the former, (N) remains comprehensible as an empirical claim in non-mathematical language: ${}_{N}$English. It is only from a mathematical point of view—or, as we might put it, in ${}_{N+}$English—that (N) 'doesn't make sense'. For the norms of mathematics to be in force and to be rigorously insisted on they need not be norms of ordinary language. They may be called 'grammatical' if that is taken to refer not to the grammar of our ordinary language, but only to a specific form of discourse, or alternatively, taking the word 'grammar' in a figurative sense, we may speak of the 'grammar', i.e. the system of rules, of a certain set of activities or of some institutionalised form of life. In a laconic remark in *Philosophical Investigations* Wittgenstein suggests that theology can be regarded as grammar (*PI* §373), providing rules for what can be said meaningfully about God. But these rules are binding only within a certain religious community. Thus, for a believer God is by definition omnipotent and benevolent.[8] To question these attributes doesn't make any sense within religious discourse: it would be 'ridiculous or blasphemous' (*AL* 32). And yet an agnostic may well do so (e.g. by arguing that if God existed he wouldn't be benevolent). You can step outside religious language, flouting its grammatical norms, while remaining within language.

Consider the following augury language-game: People have a sophisticated calculus for determining propitious days for travelling. The parameters are the number of people travelling together, their average age, and the distance to be travelled. A certain algorithm results in three numbers between 1 and 31, which specify propitious days of the month for setting

8. *Z* §717: '"You can't hear God speak to someone else, you can hear him only if you are being addressed".—That is a grammatical remark.'

off: $N * A * D = (x, y, z)$. Thus, if 3 people aged 25, 30 and 35 years respectively want to make a journey of 217 miles, they might carry out calculations resulting in the following formula:

(AC) $3 * 30 * 217 = (17, 18, 23)$.

This means that the 17th, the 18th and the 23rd would be propitious days for undertaking their journey.

Such an augury calculus would, presumably, be a kind of mathematics,[9] or at least comparable to it. For its practitioners, (AC) serves as a 'grammatical' rule, endorsing certain statements and ruling out others. The following, for example, would be ruled out by the formula as 'ungrammatical':

> (J) The 5th May would be a propitious day for 3 people aged 25, 30, and 35 years respectively to make a journey of 217 miles.

And yet, clearly, (J) is not linguistically flawed. It might even be true.

Note that for the augury calculus to have a real application, 'propitious day' must not be taken as a technical term, fully defined by the calculus. As Wittgenstein says: 'It is the use outside mathematics, and so the *meaning* of the signs, that makes the sign-game into mathematics' (*RFM* 257e). In this case, for the rule-governed transformation of three numbers into three other numbers to qualify as mathematics, the result must have a role outside mathematics, e.g. to guide us when deciding on suitable travel dates. Again, mathematics provides *additional* rules for dealing with existing concepts. Already having an idea of what a propitious travel day is, the augury calculus offers us a new method for identifying one.

These are examples of rules of 'grammar' in a wider sense of the word: not the grammar of a language like English or German, but the 'grammar' (so to speak) of a certain kind of discourse or a certain kind of prognostic activity. My suggestion is that if we follow Wittgenstein in regarding mathematical propositions as grammatical norms, we need to understand the word 'grammatical' in a similar way: not as determining what makes sense in a natural language, but rather fixing sense and nonsense in a specific kind of discourse or activity. That is, roughly speaking, an activity and discourse in which we try to develop and apply a system of *calculating* quantities, rather than simply counting or measuring them.[10]

9. But cf. *RFM* 399d.
10. Note that the comparison between mathematics and theology and augury is concerned only with the way the normativity of each of these activities is restricted and does not coincide with linguistic normativity. It is not meant to suggest that mathematics is only a matter of faith or superstition.

Note that Wittgenstein repeatedly suggests that calculations need not be laid down in sentences: that mathematics is primarily an activity, and not necessarily an entirely linguistic activity (*RFM* I §144: 93). That, too, would suggest that the norms set up by mathematics are not so much linguistic norms, as methodological rules, roughly speaking, for dealing with quantities.

> A proof is an instrument—but why do I say "an instrument of language"?
>
> Is a calculation necessarily an instrument of language, then?
>
> (*RFM* 168cd)

We can even imagine a mathematics entirely without mathematical propositions:

> People can be imagined to have an applied mathematics without any pure mathematics. They can e.g.—let us suppose—calculate the path described by certain moving bodies and predict their place at a given time. For this purpose they make use of a system of co-ordinates, of the equations of curves (*a form of description of actual movement*) and of the technique of calculating in the decimal system. The idea of a proposition of pure mathematics may be quite foreign to them.
>
> Thus these people have rules in accordance with which they transform the appropriate signs (in particular, e.g., numerals) with a view to predicting the occurrence of certain events.
>
> (*RFM* 232ab)

Wittgenstein concedes that their mathematics, which comprises no mathematical propositions, but consists entirely of practical instructions—'a technique for transforming signs for the purpose of prediction'—could hardly be regarded as 'grammar' (*RFM* 234c). For their rules are not so much rules for what one can licitly *say*, but primarily rules for what one is supposed to *do* (*RFM* 232f). Yet by 'grammar' we normally mean rules for speaking, not rules for doing.

So, to the extent that even our actual mathematics is primarily a tool for applied calculations[11] it cannot strictly speaking be called 'grammar'. But then again, there is another passage where Wittgenstein has no scruples using the term 'grammar' in an extended sense explicitly covering practical instructions:

> The grammatical rules are comparable to rules about the procedures in measuring periods of time, distances, temperatures, forces, etc.

11. MS 121, 71v: 'Mathematics consists of calculations, not of propositions.' (*Die Mathematik besteht aus Rechnungen, nicht aus Sätzen.*)

> etc. Or: these methodological rules are themselves examples of grammatical rules.
>
> (MS 117, 138–9)[12]

Again, beside such passages in which Wittgenstein uses the term 'grammar' in a wider sense, including our techniques of calculation, there are other remarks where he seems to use it in a narrower sense, formulating his view of the status of mathematics more cautiously:

> I have no right to want you to say that mathematical propositions are rules of grammar. I only have the right to say to you, "Investigate whether mathematical propositions are not rules of expression, paradigms—propositions dependent on experience but made independent of it. . .".
>
> (*LFM* 55)

> There is no doubt at all that *in certain language-games* mathematical propositions play the part of rules of description, as opposed to descriptive propositions.
>
> But that is not to say that this contrast does not shade off in all directions.
>
> (*RFM* 363cd)

> What I am saying comes to this, that mathematics is *normative*.
>
> (*RFM* 425e)

> Mathematical propositions are essentially akin to rules . . . .
>
> (*RPP I* §266)

To summarise, mathematical propositions cannot be regarded as grammatical rules if these are taken to be rules of our ordinary language (say, $_N$English or $_N$German). Rather, they are rules governing an extension of our language, or area of discourse and practice of quantification and calculation, defining a right and wrong that does not coincide with the distinction between sense and nonsense in ordinary language. An empirical prediction that is ruled out by a calculation may still be a meaningful description of a conceivable outcome, even if from a mathematical point of view it 'doesn't make sense'.

Wittgenstein's further claim that the norms of mathematics, although as such made immune from falsification, reflect our experiences may provide a first tentative answer to our question of how mathematics can be

12. *Die grammatischen Regeln sind zu vergleichen Regeln über das Vorgehn beim Messen von Zeiträumen, von Entfernungen, Temperaturen, Kräften, etc. etc. Oder auch: diese methodologischen Regeln sind selbst Beispiele grammatischer Regeln.*

such a powerful prognostic tool. But, for one thing, it seems to apply only to elementary mathematics, and for another thing, it remains to be seen how this assumed empirical basis of mathematics relates to mathematical proofs. For, as noted earlier, mathematical theorems need to be established by proof, which is obviously not the same as empirical confirmation. (We shall pursue this further in Chapters 8–10.)

It is perhaps worth noting that although Wittgenstein regards mathematical propositions as akin to grammatical rules for the non-mathematical use of certain expressions, he does not for that matter suggest that applied mathematics flows immediately from pure mathematics. It is essential to mathematics that it has an empirical application (*RFM* 257de), yet that application is never fully determined by mathematics. In fact, it is exactly because (at least a core of) mathematics has to be applicable that the meanings of its signs must to some extent be determined by their application outside maths (*RFM* 259b); just as something becomes a *logic* calculus only if its signs are interpreted to stand for expressions of ordinary language (declarative sentences, predicates, conjunctions, etc.), whose meaning is not defined, but presupposed by the calculus. Thus, Euclidian geometry treats of relations between lengths and angles, but it doesn't specify how a length or an angle is to be measured. Rather, for it to be geometry it needs to be combined with our empirical ideas of measurement (which determine our concept of length). Similarly, arithmetic doesn't give us any method of counting the number of things (*LFM* 256–7; RR 106–8). For a calculus to be arithmetic it has to be grafted onto our common-or-garden concepts of numbers of things (determined by our practice of transitive counting).

As Wittgenstein holds that the concept of a number or a length is determined by our methods of ascertaining it, he repeatedly considers the possibility of different, possibly even bizarre, methods of counting or measuring, e.g.:

> The correct method of counting might be such that after counting the first ten, say, one has to take something different as 'a further unit' (it might be a pair, say), after twenty something different still, and so on.
> (RR 108)

Or again:

> It might be practical to measure with a ruler which had the property of shrinking to, say, half its length when it was taken from this room to that. A property which would make it useless as a ruler in other circumstances.
>
> It might be practical, in certain circumstances, to leave numbers out when you were counting a set: to count them: 1, 2, 4, 5, 7, 8, 10.
> (*RFM* I §140: 91–2)

Such deviant procedures are not ruled out by mathematics. Mathematical rules determine the meaning of its signs only relative to other mathematical signs. It has obviously nothing to say about their direct empirical applicability. It teaches us how to count intransitively, that is, how to develop the series of natural numbers, but it doesn't teach us how to count apples, the techniques of transitive counting (*LFM* 258). As Wittgenstein puts it in a lecture:

> Euclidian geometry gives *rules* for the application of the words "length" and "equal length", etc. Not *all* the rules, because some of these depend on how the lengths are measured and compared.
>
> (*LFM* 256)

The basics of the meaning and 'grammar' of number words, and words such as 'length', etc., are pre-mathematical and only then developed and enriched by mathematics.

Hence, the *successful* applicability of mathematics is partly dependent on our empirical practices of counting and measuring. Different practices might require a different application of mathematical propositions (RR 123), or indeed a different calculus: If, for example, we counted as Wittgenstein imagines we might, in the passage quoted above (from RR 108), it might be more useful to employ (an arithmetic that yields) the equation '$6 + 6 = 11$'. So, mathematical proof alone cannot guarantee that a theorem is suitable to be used as a norm of expression. Its status as norm of expression is only contingent, even though a proof has given it the dignity of a necessary truth.

Once we move beyond rather elementary areas of mathematics, the link with applications in which mathematical theorems are supposed to play a normative role becomes more problematic. That is fairly obvious in areas of pure mathematics that are developed without any application in mind, or not the one that is ultimately found for them.

About 300 B.C. Euclid proved that there is no greatest prime number. Very large prime numbers play a role in cryptography, but it is hard to see how the proven infinity of primes could be applied outside mathematics, in accordance with Wittgenstein's idea of mathematical normativity.

Euclid also established that for every *Mersenne prime* number (i.e. a prime number of the form $2^p - 1$, where p is itself a prime) there is a *perfect* number (i.e. a number that is the sum of its proper factors), namely: $P = 2^{p-1} (2^p - 1)$. And in the 18th century Euler proved that every even perfect number can in this way be derived from a Mersenne prime, so that there is a one-to-one correlation between the two kinds of numbers (Higgins 2011, 28). However, neither Euclid nor Euler were concerned to think of any applications for their discoveries, and, as far as I know, none have been found yet.

In 1854 in his inaugural lecture on geometry, Georg Bernhard Riemann was trying to show that Euclid's geometrical axioms were empirical, rather than self-evident truths. He suggested an alternative, but equally consistent axiomatic system of geometry, according to which space would appear unbounded, rather than infinite. This non-Euclidian, double-elliptic geometry (in which there are *no* parallel lines and the sum of the angles of a triangle is always *greater* than 180°) was not initially intended by Riemann to rival Euclidian geometry in its application to three-dimensional physical space, but merely to demonstrate the empirical status of Euclid's axioms. The only application for double-elliptic geometry that Riemann, and a little later Eugenio Beltrami, thought of was the surface of a sphere, in which case, however, 'straight lines' would not actually be straight, as a ruler's edge, but great circles (whose centre is the centre of the sphere). Only half a century later Riemann's geometry was applied to cosmic space, as an alternative to Euclidian geometry, in Einstein's theory of general relativity (Kline 1980, 85–6, 180–1).

In all these cases, the proof of a theorem does *not* establish a norm of expression,[13] but only, as we might say, a *candidate* for a norm of expression in empirical discourse. Wittgenstein acknowledged that the realm of mathematics is greater than that of applied or applicable mathematics. He realized that new branches of mathematics are developed 'without reference to any possible application at all' (RR 132). At least for the time being such constructions can be regarded as 'systems of "empty connexions"':

> It may be that applications of these new branches of mathematics will be found, perhaps quite shortly, perhaps only after many years; that has happened, of course. Or it may be that no application will ever be found. Again, if some new branch of mathematics is given an application, this may be in some rather unexpected way.
>
> (RR 132)

This adds further to the reasons given above not to regard mathematical propositions as ordinary grammatical propositions. I argued that well-established mathematics (basic arithmetic and Euclidian geometry) provides norms of representation not for ordinary language, but for its extensions, some of them only familiar to a relatively small number of mathematically educated professionals. Beyond that, when new theorems are produced in mathematical research they often remain in search of an application for some time: in search of suitable new concepts in a scientific theory to fit those theorems. So for some time they can only be regarded as *candidates* for norms of expression.

13. Except for descriptions of mathematicians' activities (cf. *LFM* 47).

7 Rule-Following

As argued in the previous chapter, Wittgenstein's claim that mathematical propositions are rules of *grammar* needs some qualification. It would only be the grammar of an extension of our language, for the most part familiar only to specialists. Alternatively, regarding mathematics as a practice of techniques of calculation, rather than a part of language, we may say that the word 'grammar' has to be taken in a wider, perhaps metaphorical sense, and even then we have to remember that the applicability of new pieces of mathematics is often only a future possibility; hence such new parts of pure mathematics are, for the time being, only *candidates* for rules of 'grammar' in the envisaged sense. Even so, mathematics can undoubtedly be said to be based on rules. Yet Wittgenstein has famously alerted us to a philosophical problem about following rules, which can be presented as follows:

> "But how can a rule teach me what I have to do at *this* point? After all, whatever I do can, by some interpretation, be made compatible with the rule."
>
> (*PI* §198)

He illustrates this problem with a school scenario in which a stubbornly deviant pupil is taught to write down the series of even numbers (on the command '+ 2'), does so correctly up to 1000, but then continues: 1004, 1008, 1012.

> We say to him: "Look what you've done!"—He doesn't understand. We say: "You were meant to add *two*: look how you began the series!"—He answers: "Yes, isn't it right? I thought that was how I was *meant* to do it."
>
> (*PI* §185)

What is the point of imagining this curious case?—The wording of §3 of *RFM* I (where the case is alluded to) gives us a hint: '*How do I know* that in working out the series + 2 I must write "20004, 20006" and not

"20004, 20008"?' Before, the key term was 'determine': How can a formula determine certain steps? The epistemological paraphrase 'How do I know given a formula what steps to take?' makes it clearer what kind of determination the question is looking for: The underlying idea is that a formula *determines* certain steps by *telling* us what steps to take. The formula is to contain the information, to give us the knowledge as to what steps to take. Elsewhere Wittgenstein makes the idea of determination through knowledge explicit:

> What must I know in order to be able to carry out the order? Is there some *knowledge* that makes the rule followable only in *this* way?
> (*RFM* 341h)

But then, as the scenario of the deviant pupil illustrates, the formula doesn't appear to contain the information required. And indeed it is hard to see how it could. After all, the series + 2 contains infinitely many steps; yet how could a laconic order or formula contain an infinite amount of knowledge?

The second paragraph of *RFM* I §3 attempts an answer. It is only one piece of information that the formula needs to convey, namely the principle of adding 2. Once you have understood that, you know that it involves only a small number of modifications from one number to the next in the series: In the units you change 0 to 2, 2 to 4, 4 to 6, 6 to 8, or 8 to 0, and whenever you make a change to 0 you also move one up in the next digit to the left. Wittgenstein's reply is that this doesn't help, as the philosophical problem applies even to the simplest algorithm, even to the instruction always to repeat the same number: 2, 2, 2, 2 . . . 'For how do I know that I am to write "2" after the five hundredth "2"?' We may want to reply that, surely, the instruction is clear enough: we just keep writing the *same* figure. But Wittgenstein objects that the problem is to know what at any given point *counts* as 'the same figure'.

In a parenthesis at the end of the first paragraph he compares that question to the question 'How do I know that this colour is "red"?'. Again, one could imagine the reply that 'red' is what we call whatever is of the same colour as a given sample of red, e.g. a ripe tomato. But Wittgenstein's concern would be that we'd still need to know what counts as 'of the same colour' in a given situation. And we can imagine our criteria for colour identity to be less than straightforward. They might, for example, involve a reference to the time of day, so that what counts as 'red' in the morning, say, would not be called 'red' at night (cf. *RFM* 328f–h).

This shows the wide range of the problem of rule-following and why it plays a prominent role not only in Wittgenstein's philosophy of mathematics, but also in his philosophy of language and mind, as presented in

Philosophical Investigations. An instruction rule (such as "keep adding 2") can be variously understood. We may make it more explicit, complementing the original instruction with second-order instructions of how to interpret it (cf. *PI* §86), third-order instructions as to how to interpret the second-order instructions, and so forth. Yet such additional explanations cannot resolve the philosophical problem, since *any* verbal or symbolic expression can, again, be interpreted in different ways. Ultimately, the problem of rule-following is the *problem of general terms*. How can we assure our understanding of a word with an open-ended number of applications?[1] We agree on a range of instances of a concept *F* (e.g. '+ *2*', 'red') and that it is to be applied in the same way in all future cases. But what exactly will count as 'the same' (cf. *PI* §225)?[2]

The possible misunderstandings Wittgenstein considers may sound far-fetched, but they highlight the point that, logically speaking, there always remains a gap between a general instruction (applicable to an indefinite number of instances) and its execution. That is to say that although, of course, in some sense an instruction can be said to determine some answers as correct and others as incorrect (*RFM* I §1: 35),[3] this determination is not foolproof. Wittgenstein sets his face against a certain philosophical picture of such a determination according to which the correct

1. This paraphrase shows that, in the relevant sense, *all* words are general, for all words—even proper names—are to be used on indefinitely many occasions.
2. Note, therefore, that Wittgenstein's concern for this problem in the *Philosophical Investigations* is not based on the assumption that language is governed by rules. For one thing, the rules discussed in the rule-following considerations (*PI* §§185–242) are not semantic rules (definitions), but instruction rules: open-ended orders. For another thing, although at one time Wittgenstein was inclined to regard language as something like a rule-governed calculus, in his later writings he repudiated this view: He continued to stress the normativity of language, but at the same time he realised that linguistic normativity was typically piecemeal: not determined by general rules, but by our ability to acquire relatively stable patterns of linguistic usage without the guidance of general rules or definitions. Hence the philosopher's quest for definitions (semantic rules) is answered negatively, by the notion of family-resemblance concepts (*PI* §§65–74). So, the universal relevance of the rule-following discussion is not due to the questionable assumption that language is everywhere governed by rules; rather: following rules is just a particularly straightforward manifestation of linguistic understanding: of the open-ended understanding of general terms (cf. Schroeder 2017b, 258–62).
3. That Wittgenstein is not promoting any form of 'scepticism' about rule-following is also brought out nicely in the following remark from a lecture:

> It is felt to be a difficulty that a rule should be given in signs which do not themselves contain their use, so that a gap exists between a rule and its application. But this is not a problem but a mental cramp. . . . We are only troubled when we look at a rule in a particularly queer way.
>
> (*AL* 90)

answer is somehow already *contained* in the instruction. This was a view he found in Frege and Russell:

> In his fundamental law Russell seems to be saying of a proposition: "It already follows—all I still have to do is, to infer it". Thus Frege somewhere says that the straight line which connects any two points is really already there before we draw it; and it is the same when we say that the transitions, say in the series + 2, have already been made before we make them orally or in writing—as it were tracing them.
>
> (*RFM* I §21: 45)[4]

Wittgenstein insists that that is only a picture, a metaphor (*RFM* I §22: 45). Admittedly, giving someone a formula by which to calculate a series of 50 numbers can be just as reliable as asking him to copy an existing list of those numbers; nevertheless the former case is *not* one of copying or tracing what is already there. What *can* be produced does not, for that matter, exist already as a shadow in some Platonic underworld; not really (cf. *PG* 281d).

And even if it did—: 'how would it help?' (*PI* §219). Even if the case were one of copying, we could still imagine a pupil doing it in an aberrant way—for instance, leaving out every tenth number (cf. *PI* §86)—, thinking that was the correct method.

The correct answers do no follow mechanically from the rule or formula. Whenever we take the step from the general rule to an individual application, we have to *take* the rule in a certain way. Hence one could say that following a rule 'always involves interpretation' (*RFM* I §114: 80). Not, to be sure, in the substantive sense of verbally giving it an interpretation: 'substituting one expression of the rule for another' (*PI* §201), but in the minimal sense of *applying the rule in one way* rather than another.

At one point, in 1929, Wittgenstein was inclined to think that whenever we proceed to apply a general rule in an individual case—as the latter is not actually already contained in the former, but requires us to go *beyond* it—we need a new insight or intuition:

> Supposing there to be a certain general rule (therefore one containing a variable), I must recognize each time afresh that this rule may be applied here. No act of foresight can absolve me from this act of insight. Since the form to which the rule is applied is in fact different at every step.
>
> (*PR* 171)

4. Cf. Frege 1884, 24: 'Each [axiom of geometry] contains concentrated within it a whole series of deductions for future use'; cf. 99–103. Cf. *LFM* 144–5; *PI* §§218–19.

And, as he later explained in a lecture, he took this to be a fundamental tenet of Intuitionism, L.E.G. Brouwer's position in the early 20th-century debate about the foundations of mathematics:

> Intuitionism comes to saying that you can make a new rule at each point. It requires that we have an intuition at each step in calculation, at each application of a rule; for how can we tell how a rule which has been used for fourteen steps applies at the fifteenth?—And they go on to say that the series of cardinal numbers is known to us by a ground-intuition—that is, we know at each step what the operation of adding 1 will give. We might as well say that we need, not an intuition at each step, but a decision.
>
> (*LFM* 237)

However, he soon came to reject this view. 'If intuition is needed to continue the series + 1, then it is also needed to continue the series + 0' (*RFM* I §3c: 36; cf. *PI* §214); but it isn't. Returning to his earlier remark, quoted above: 'No act of foresight can absolve me from this act of insight [for every new step]', he wrote in the margin: 'Act of *decision*, not of *insight*' (*PR* 171). The same point is elaborated in *The Brown Book*:

> It is no act of insight, intuition, which makes us use the rule as we do at the particular point of the series. It would be less confusing to call it an act of decision, though this too is misleading, for nothing like an act of decision must take place, but possibly just an act of writing or speaking. . . . *We need to have no reason to follow the rule as we do*. The chain of reasons has an end.
>
> (*BB* 143; cf. *PI* §186)

To know something 'by intuition' means: to know it immediately, without reasoning, as if by an act of direct perception. You just *see* that something is the case, though not literally, but with your mind's eye. (Brouwer also speaks of 'introspection'.) However, you can only have a (true) intuition that something is the case, if indeed it is the case. The object of an intuition must be an objective fact and hence ascertainable independently of someone's intuition; just as the object of one's visual experience—if it is not a hallucination—must have an independent existence that can be ascertained in other ways (e.g. by touch). Normally, when we say that somebody had an intuition that something is the case, we know of another, more pedestrian way of recognizing the matter. Thus, you can be said to know by intuition that $27 \times 177 = 4779$ if you can immediately produce the answer that others derive by calculation (*LFM* 30). In short, what is known by intuition—without reasoning (or evidence, or sense perception)—must also be ascertainable by reasoning (or on the basis of evidence or through sense perception). In the case in hand, however, it is not only that an answer is given immediately and without reasoning; the important point is that ultimately no reason *can* be given. That is to say, the reason why you write 2002 after 2000 is of course: that you are applying the rule + 2; but when you are questioned further: why when applying the rule + 2 you write 2002 after 2000, you can only say that that is what adding 2 requires at this point. Your reasons soon give out; and then you act without reasons (*PI* §211). It is not only an immediate apprehension of something that can also be established or justified by reasons; in the end there just are no more reasons. And that means that the concept of an intuition (although tempting to invoke) is out of place.

Wittgenstein rejects the talk of intuitions of how to continue a series as just another version of the Frege-Russell idea that the steps to be taken are, in some shadowy sense, already taken. In order for something to be seen by intuition it must already be there. In this respect the word 'intuition' is akin to 'discovery' (*LFM* 82).

As an antidote to the idea of an intuition guiding our steps in following a rule, Wittgenstein suggests that it would be closer to the mark to speak of a *decision*. To say that I have to *decide* to write 2002 after 2000 (in following the rule + 2) makes it clear that the number cannot be read off anywhere: that the application of the rule does not already exist in some Platonic realm. Rather, I have to take full responsibility myself for producing it. And yet, it is misleading to speak of a decision, for that might wrongly suggest that feeling free to go for whatever number I choose, I just happen to pick 2002. It is not like that, of course. In fact, in following the rule I feel *compelled* to write this number (*RFM* 413d, *PI* §231). 'When I obey a rule, I do not choose' (*PI* §219; cf. *LFM* 237–8).[5]

There is a danger when reading Wittgenstein to mistake his rejection of a natural philosophical *explanation* of a given phenomenon for a rejection of the *phenomenon* itself. It is tempting to think of following a rule as tracing steps that, in some way, have already been taken—be it objectively, in some Platonic realm, or subjectively, in some mysterious act of meaning. This way of thinking of rules can be so deeply rooted in one's mind that Wittgenstein's objections to it sound like an attack on the very possibility of following a rule. Thus it can appear that if in following a rule the step to be taken at a given point cannot be intuited, cannot be perceived—it doesn't exist. It would appear that there is no *correct* answer to the question of how to apply the rule (which would of course

5. *Pace* Dummett who wrongly attributes to Wittgenstein an unbridled decisionist position: 'at each step we are free to choose to accept or reject the proof. . . . In doing this we are making a new decision. . . . [According to Wittgenstein, we have] freedom of choice in this matter'. 'He appears to hold that it is up to us to decide to regard any statement we happen to pick on as holding necessarily, if we choose to do so' (1959, 495–6, 500).

mean that there was no such thing as following a rule). And this radically destructive reading would seem to be supported by Wittgenstein's use of the word 'decision' in this context, especially if one doesn't pay attention to the exact wording, which presents it as right in only one respect, but wrong in another.

The continuation of the passage from the 1939 lectures quoted above makes it quite clear that Wittgenstein rejects that 'intuitionist' view that afterwards has often been ascribed to him, according to which a rule can never really tell us what to do at a given step:

> Intuitionism comes to saying that you can make a new rule at each point. It requires that we have an intuition at each step in calculation. . . . We might as well say that we need, not an intuition at each step, but a decision.—Actually there is neither. You don't make a decision: you simply do a certain thing. It is a question of a certain practice.
>
> Intuitionism is all bosh—entirely.
>
> (*LFM* 237)

Again, the fourth paragraph of *RFM* I §3 makes it clear too that Wittgenstein had no such absurdly radical consequences in mind. It is perhaps the most elegant account of (the core of) Wittgenstein's solution to the rule-following puzzle:

> "But do you mean to say that the expression '+ 2' leaves you in doubt what you are to do e.g. after 2004?"—No; I answer "2006" without hesitation. But for that very reason it is superfluous to suppose that this was determined earlier on. My having no doubt in face of the question does *not* mean that it has been answered in advance.
>
> (*RFM* I §3d: 37)

Our initial inclination was to think that in a straightforward case of rule-following, such as + 2, where there is never any doubt about the next step, it must have been determined (i.e. this particular step must have been taken somewhere) in advance:

(1) Transition is not doubtful ⇨ transition has been determined in advance.

For it seemed that in order to be certain of a transition at a given point we must somehow be able to ascertain that that particular transition has been laid down as correct. When more careful consideration shows that we don't in fact find the correct answer laid down in advance, one may be inclined to argue from (1) by modus tollens that we cannot be certain of it. That is the response given by the first sentence of *RFM* I §3d. Yet Wittgenstein insists on the truth of the antecedent (which it would be absurd to deny), but rejects his interlocutor's conditional, suggesting a different one instead:

(2) Transition is not doubtful ⇨ transition need not be determined in advance.

That of course raises the question of how we can be so certain of the application of a rule at a given point if that application has not been laid down anywhere in advance. The answer was already given in *RFM* I §1a and is repeated more explicitly in *RFM* I §22: There is an indirect way of determining the development of a series, without specifying each individual step in advance, namely through training. Thus, for instance, children get training 'in the multiplication tables and in multiplying, so that all who are so trained do random multiplications (not previously done in the course of being taught) in the same way and with results that agree'; and the same holds for the series + 2 (*RFM* I §22: 45–6).

As noted in the first section of *RFM* I, a formula *can* determine a series of transitions, but only for those who have been trained to use it in a certain way. And this training provides a determination that is not mediated by any theoretical knowledge. Having mastered the technique of adding 2, I know at each point of the series what to write next. But the question '*How* do you know?' (from the beginning of *RFM* I §3) cannot be answered. My knowledge is a practical certainty, based on training, not based on reasons. In the same way, our mastery of colour words is not mediated by theoretical knowledge (*RFM* I §3a: 36). Why do I call that colour 'red'? I cannot give a reason, I can only cite a cause: 'I have learned English' (*PI* §381).

Returning to our original statement of the rule-following paradox:

> "But how can a rule teach me what I have to do at *this* point? After all, whatever I do can, by some interpretation, be made compatible with the rule."
>
> (*PI* §198)

—it contains the fallacious inference:

> A rule *can* be interpreted in various ways.
> Therefore, a rule *cannot* determine a continuation.

One might as well argue:

> Your bicycle could always be stolen.
> Therefore, it can never be used.

The reply is, of course: *if* it is stolen from you, then you cannot use it; but it is well possible, perhaps even likely, that it is not stolen and then you *can* use it. Similarly: for any given continuation (e.g. 1000, 1002, 1004), a suitable rule (e.g. $x_n = 2n$) can always be interpreted *not* to yield that result—but it need not be so interpreted and normally it isn't. Roughly speaking, it is fallacious to argue from 'It can go wrong' to 'It can't go right'. Moreover, even if your bicycle is stolen, it can (and probably will) still be used—though not by you. Likewise: if the rule '$x_n = 2n$' is interpreted in a deviant way, it will still determine a continuation of the series, albeit not the one we expected (but perhaps: 1000, 1004, 1008).

Perhaps the worry is that the rule needs to be interpreted at all. It cannot *by itself* determine a continuation. To understand it in some way, it seems, you need to give it an interpretation. And once you accept that an interpretation is necessary, you seem to be launched on an infinite regress: for whatever interpretation you give, it is in no better a position than the rule itself. It, too, needs to be interpreted in a particular way (cf. *PI* §86)—and so on, and so forth.

Let us first consider the idea that a rule, or formula, cannot by itself determine a continuation. Wittgenstein's reply is that this is a misuse of language that would result in the abolition of an important distinction (*PI* §189). There are formulae (like '$y \neq x^2$') that do *not* determine the value of y for a given x; other formulae (such as '$y = x^2$') *do* determine the value of y for any x. It would be foolish to ignore this difference.—But this will not satisfy Wittgenstein's interlocutor, who may retort: 'True, one type of formula does, while the other does not determine a value for a given number; but not *by itself*. $y = 3^2$, for instance, yields 9 *only when interpreted in a certain way*. Thus, strictly speaking, it is not the formula that determines the result, but the formula plus a certain interpretation. And now we are back with the infinite regress problem: for the interpretation would itself be in need of an interpretation, and so on.'—Wittgenstein has this reply:

> It can be seen that there is a misunderstanding here from the mere fact that in the course of our argument we give one interpretation after another; as if each one contented us at least for a moment, until we thought of yet another standing behind it. What this shews is that there is a way of grasping a rule which is *not* an *interpretation*, but which is exhibited in what we call "obeying the rule" and "going against it" in actual cases.
>
> (*PI* §201)

For us to think, even for a moment, that on a certain understanding the formula '$y = x^2$' yields 9 for $x = 3$, the infinite regress that seemed to threaten must have been stopped. Our understanding in this case cannot just be an interpretation: that is, another formula to paraphrase the first, of which again it would be an open question how to understand it. Our understanding of a rule need not be mediated by another rule. What it amounts to is simply this: we are able to work out that $x = 3$

yields 9. How? Well, 'I have been trained to react to this sign in a particular way, and now I do so react to it' (*PI* §198). This is the core of the matter: know-how (a skill) cannot, ultimately, be explained in terms of know-that (a piece of information). For any piece of information, for any formula in the mind, we would again need to *know how* to apply it. *Occasionally*, our understanding of a formula may be based on, or mediated by, our understanding of another formula. Thus, when we learn to read a new notation, e.g. '*x!2*' as 'x^2' (*PI* §190). But such a translation of one formula into another cannot be the paradigmatic case of understanding, certainly not the basic case of understanding: or else we would never *begin* to learn the meaning of any formula. Likewise, our basic linguistic understanding cannot be accounted for in terms of translation into another language. This may be appropriate for one's understanding of some *foreign* words or phrases, but obviously not for the initial mastery of one's mother tongue (cf. *PI* §32).

The suggestion under discussion was that a fairly elementary mathematical capacity (such as developing the series of even numbers) could be explained as the result of having grasped a rule; that the ultimate foundation of mathematical competence was explicit rule-following. Now it turns out that it is the other way round. Far from providing a theoretical foundation for basic know-how, explicit rule-following does itself presuppose some know-how for which no further foundation or justification is possible. It is simply the way we continue after some suitable training, 'by means of *examples* and by *practice*' (*PI* §208).

That may sound hard to accept. After all, the examples run through in training and practice are limited and do not logically entail any particular continuation. One may want to protest:

> How can he *know* how he is to continue a pattern by himself—whatever instruction you give him?—Well, how do I know?—If that means "Have I reasons?" the answer is: my reasons will soon give out. And then I shall act, without reasons.
>
> (*PI* §211)

Again and again, we are strongly inclined to insist that somehow all the correct answers must be laid down in advance in the mind, so that each step can be justified by reference to that perfect mental instruction manual. But that perfect mental instruction manual is an illusion and (as explained) a logical impossibility. Our basic skills must stand on their own. They are engendered by training, but what they enable us to do is open-ended and cannot be exhaustively listed in training, nor could there be such an exhaustive list in the mind or anywhere else.

> With the words "*This* number is the right continuation of this series" I may bring it about that in the future someone calls such-and-such the "right continuation". What 'such-and-such' is I can only show

> in examples. That is, I teach him to continue a series (basic series), without using any expression of the 'law of the series'. . . .
>
> He must go on like this *without a reason*. Not, however, because he cannot yet grasp the reason but because—in *this* system—there is no reason. ("The chain of reasons comes to an end.")
>
> . . . For at *this* level the expression of the rule is explained by the value, not the value by the rule.
>
> (*Z* §§300–1)

***

7.1 Rule-Following and Community

It is still widely believed that according to Wittgenstein rule-following must be a *social* practice: that it requires the agreement of a community to constitute standards of correctness (e.g. Frascolla 1994, 111ff.). That is definitely a misinterpretation.

Note first of all that recourse to a community provides no solution to Wittgenstein's paradox of rule-following. The paradox results from the natural assumption that what we mean or understand (by a rule, an order, or a word) must be absolutely determined by some mental representation together with the observation that in fact it cannot be so determined. Can it be determined if we replace, or complement, our mental acts of meaning or understanding by the responses of other members of our linguistic community? No. Consider again the case of the deviant pupil (*PI* §185): What comes after 1000 in the series of + 2? If the formula '*1000 + 2*' cannot determine 1002 as the correct result, neither can the words of a bystander. If the teacher's 'add 2' can be systematically misunderstood (to mean: 'add 2 up 1000 and then add 4', for example), then similarly, the teacher's (or other people's) comment 'That's right' can be understood to mean 'That's false' when applied to calculations above 1000. Or again, the pupil may think that in the case of four-digit numbers, what he writes is always to be 2 above what others say. (This would be comparable to the way others' use of the pronoun 'you' when talking to him is in agreement with what he says using the different word 'I': '*You* go home'.—'Yes, *I* go home'.) If, when the teacher complains that what the pupil wrote at a given point in the series is different from what the teacher and others wrote, the pupil replies: ' "Different?—But this surely *isn't* different!"—what will you do?' (*RFM* I §115: 80). The deviant pupil's manoeuvre applies not only to rule formulations but also other people's responses, for instance, to the way his teacher and others urge him to understand those signs:

> "But if you want to remain in accord with the rules you *must* go this way."—Not at all, I call *this* 'accord'.—"Then you have changed the

> meaning of the word ‘accord’, or the meaning of the rule.”—No;—who says what ‘change’ and ‘remaining the same’ mean here?
>
> (*RFM* I §113: 79)

As far as Wittgenstein’s puzzle about rule-following is concerned, communal agreement is a red herring. As long as we hold on to the philosophical misconception of understanding as in some sense *containing* its object, a formula is just as insufficient as the responses of others for providing that—chimerical—form of understanding. But once we have freed ourselves from the philosophical picture of understanding that Wittgenstein attacks, we find that there is no longer any problem in the idea of a formula determining a series; it has a perfectly ordinary sense (*RFM* I §1: 35); and there is no need to invoke the agreement among a community of speakers.

Hence it is not surprising that in the *Philosophical Investigations* (which contains the most carefully edited account of Wittgenstein’s thoughts on rule-following) there is no textual support for the view that rule-following requires a community. In *Investigations* §199 Wittgenstein makes the point that rule-governed forms of behaviour must be a recurrent practice or custom: ‘It is not possible that there should have been only one occasion on which someone obeyed a rule. It is not possible that there should have been only one occasion on which a report was made, an order given or understood; and so on’ (*PI* §199). There is no implication that a solitary person could not have a practice of following a rule. Of course that would be an exceptional case. For as a matter of fact, human beings do not live in complete isolation, and our most important and sophisticated normative practices (language, mathematics) have developed from social interaction and are often forms of social interaction (making a report, giving an order). Hence the most natural examples of rule-governed forms of behaviour, taken from our everyday life, will of course tend to have a social background and involve more than one person. Shared rules are what we are mostly interested in, but they are not the only ones that are possible.

The most likely passage to be cited from the *Investigations* in support of the view that Wittgenstein regarded rule-following as essentially a *social* practice is this:

> And hence [because there is a way of grasping a rule which is exhibited in correct applications] ‘following a rule’ is a practice. And to *think* one is following a rule is not to follow a rule. Hence it is not possible to follow a rule ‘privately’: otherwise thinking one was following a rule would be the same thing as following it.
>
> (*PI* §202)

However, it is significant that the word ‘privately’ occurs in inverted commas. That is because it does not simply mean: in private, as is made clear by the final clause: 'privately' means: in a way that there can be no distinction between following a rule correctly and only believing to follow a rule. Yet even in the privacy of my study, apart from any community, I can draw that distinction: I can, for example, write a diary in a secret code (as Samuel Pepys did) and from time to time consult a table to check if what I think is the sign for a given letter is in fact the correct sign. So this passage does not exclude solitary rule-following. Rather, the 'privacy' that is ruled out here (because it does not allow for the distinction between appearance and reality) is the 'privacy' of one's mind where no independent check of one's impressions is possible. That is confirmed in *Investigations* §256, where Wittgenstein calls a language 'private' (in inverted commas) only if nobody else could *possibly* understand it. Hence something like Pepys's secret code would not be a 'private' rule, in Wittgenstein's sense, as even if nobody else happens to know it, it is well possible for others to learn to understand it and then to check if it was applied correctly.

The closest Wittgenstein gets to the community thesis is in MS 164, a first draft notebook, later published as Part VI of *Remarks on the Foundations of Mathematics*. There is in particular one pertinent passage where Wittgenstein *raises the question* whether a solitary person could calculate or follow a rule, or whether like trade any rule-following requires at least two people (*RFM* 349ef)—but no answer is given and the matter is not pursued any further. In other manuscript remarks the question is answered in no uncertain terms: even a completely isolated individual, like Robinson Crusoe, could use a language if his behaviour was sufficiently complex (MS 124, 213 & 221; MS 116, 117). Anyway, none of this found its way into the *Investigations*. Obviously, Wittgenstein did not in the end attach all that much importance to the matter.

Admittedly, rule-following or the application of predicates require some standard of correctness. But such a standard need not be supplied by other people, it can also consist in a written table which I consult when in doubt. (And if now you want to object that the written table can be misunderstood, the answer is, as illustrated earlier: so can other people.) Indeed, even independent applications of the same sign or rule by a single speaker can confirm each other and thus allow for a distinction between right and wrong (cf. Schroeder 2001, 189–91). After all, it is not pointless to check and double-check your own calculations. Contrary to what the champions of the community thesis want to make us believe, we can and often do correct our impressions by comparison with our own impressions at other times. Indeed, logically it makes no difference whether I check my calculation again and again (checking it against my own continuing arithmetic practice) or whether I ask you to check it.

Imagine a desert islander: Is there any reason why he should not be able to carve some symbolic notches into the wall of his hut to indicate the number of coconuts he has in store? In this case the standard of correctness is particularly straightforward: he can always check if for every notch there is indeed a coconut. Even defenders of the community thesis tend to shrink from the implausible consequence that a Robinson Crusoe could not be said to follow a rule. 'I do not see that this follows', writes S. Kripke: 'What does follow is that *if* we think of Crusoe as following rules, we are taking him into our community and applying our criteria for rule following to him' (1982, 110).[6] But then the community thesis becomes quite vacuous, as now there does not seem to be anything it rules out. What it boils down to is that if *we* judge or imagine somebody to follow a rule, then we apply *our* criteria for rule-following. Well, obviously. That is true of all predications: if we call something a tree then we apply our criteria for something's being a tree; but it does not follow that there could not be any tree in our absence.

Rule-following or the use of concepts need not be a *social* practice, but of course with us they usually are. Natural languages as well as mathematics are used by large communities of people for their communication and cooperation. Clearly, for the semantic norms of English and the rules of arithmetic communal agreement is required and well-established:

> Disputes do not break out (among mathematicians, say) over the question whether proceedings have been in accordance with a rule or not. People don't come to blows over it, for example. This belongs to the scaffolding from which our language operates (e.g. gives descriptions).
>
> (*PI* §240)

But if (as demonstrated) correctness cannot ultimately be based on a formula or definition, nor on any mental occurrences of meaning or understanding, must correctness in such cases not after all be a matter of communal agreement? If rule-following is essentially a practice (*PI* §202), then following communal rules must be a social practice, whose standards of correctness, it would appear, are determined by the community. That is the response Wittgenstein expects his readers to make:

> "So you are saying that human agreement decides what is true and what is false?"
>
> (*PI* §241)

And that is indeed what some readers seem to have taken him to say. Thus, Pasquale Frascolla held it to be 'Wittgenstein's pivotal thesis that the ultimate source of legitimacy for mathematical concepts and proofs is

6. In a similar vein: Peacocke 1981, 93; von Savigny 1996, 122.

communal ratification' (Frascolla 2001, 284). Wittgenstein invites us to consider what that would look like in the case of a colour word:

> You say "*That* is red", but how is it decided if you are right? Doesn't human agreement decide?—But do I appeal to this agreement in my judgments of colour? Then is what goes on like *this*: I get a number of people to look at an object; to each of them there occurs one of a certain group of words (what are called the "names of colours"); if the word "red" occurred to the majority of the spectators (I myself need not belong to this majority) the predicate "red" belongs to the object by rights. Such a technique might have its importance.
>
> (*Z* §429)

Is that the way we use colour words? Of course not. It is rather like this:

> *Colour-words* are explained like *this*: "That's red" e.g. [pointing at something red]. Our language game only works, of course, when a certain agreement prevails, but the concept of agreement does not enter into the language-game. If agreement were universal, we should be quite unacquainted with the concept of it.
>
> (*Z* §430)

Similarly, in mathematics: it is not our language game to vote on the truth of a mathematical proposition or proof. There is no such thing as 'communal ratification' in mathematics. The inclination of the majority does not count as a legitimate reason for accepting a mathematical proposition as true. As Wittgenstein points out, we have to distinguish between an *external* and an *internal* perspective: agreement is, obviously, important as a prerequisite of the shared language game, but plays no role *in* it (*RFM* 365g). For our communal mathematical culture it is important that at the point where our concepts cannot be explained further, but are rooted in practice, we agree. But this requirement plays no cognitive or normative role. For, unlike football or theatre, mathematics is not intrinsically social: at no point do the rules of mathematics refer to, or require the existence of, other people. We need our basic inclinations to be in sync, but at the same time they must all be independent. We must all find that the next even number after 1000 is 1002, but we must do so all of us independently, without having to consult each other. 1002 *is* the correct answer, but not because it is the general view. We don't use people's agreement as a criterion of correctness. 'What criterion do you use, then? None at all' (*RFM* 406c).

8 Conventionalism

According to Wittgenstein, mathematical propositions are (something like) rules of grammar, that is, it would appear, conventions (*AL* 156–7, *BT* 196, *RFM* 199a, TS 309, 89; cf. *PG* 190, MS 108, 97), or at least determined by conventions. A mathematical proposition doesn't describe a fact (*RFM* 356ef), but serves as a linguistic convention: 'only supposed to supply a framework for a description' (*RFM* 356f), determining the correct use of language (*RFM* 165h, 196f): what in a certain area of discourse makes sense and what doesn't (*RFM* 164bc). To that extent it appears natural to regard Wittgenstein's position as a form of conventionalism.

Perhaps the best-known presentation of conventionalism in the philosophy of mathematics is due to the Logical Positivists, in particular A.J. Ayer, who defended the view that all mathematical truths are analytic (1936). That is to say, they can be derived from a set of conventions defining the meanings of our mathematical symbols. For instance, the conventional definition of the series of natural numbers in terms of addition of 1 (each number > *1* being defined as its predecessor + *1*) logically implies any correct equation of the form $a + b = c$. Thus, Ayer argued, an equation such as '*7* + *5* = *12*'—far from being, as Kant thought, synthetic a priori—can be proven by a succession of definitional substitutions (Ayer 1936, 109–15).[1]

Gordon Baker and Peter Hacker have argued, however, that Wittgenstein's view of necessary truths was quite different from the logical empiricist position. On Wittgenstein's view, Baker and Hacker insist, necessary truths do not *follow from* semantic conventions, but rather are partly *constitutive of* the meaning of their constituent expressions. To regard necessary truths as consequences of semantic conventions is to be guilty of the 'meaning-body' confusion, which Wittgenstein attacked in the early 1930s (Baker & Hacker 2009, 367–70)—:

> Are the rules, for example, $\sim\sim p = p$ for negation, responsible to the meaning of a word? No. The rules constitute the meaning, and are

1. For a discussion of this claim see Chapter 9.

> not responsible to it. . . . Rules are arbitrary in the sense that they are not responsible to some meaning the word already has. If someone says the rules of negation are not arbitrary because negation could not be such that $\sim\sim p = \sim p$, all that could be meant is that the latter rule would not correspond to the English word "negation". The objection that the rules are not arbitrary comes from the feeling that they are responsible to the meaning. But how is the meaning of "negation" defined, if not by the rules? $\sim\sim p = p$ does not follow from the meaning of "not" but constitutes it.
>
> (*AL* 4; cf. *PG* 52–3; 184; *LFM* 282; *RFM* 106b)

What Wittgenstein is concerned to undermine here is the idea of the meaning of a word as 'something over and above the use of the word'; some abstract or psychological entity 'attaching to the word itself' (*RFM* I §13: 42), from which the rules for its use could be derived and against which they could be checked. To use a chess analogy: one may be tempted to think that each piece has an inner nature that determines its possible moves, and that the rules of chess have to be derived from, and must be in accord with. The truth is, of course, that the rules of the game were not derived from anything, but are arbitrary stipulations; and the rôle of a piece in the game is determined, or constituted, by the rules.

However, the fact that the chess rules cannot be derived from some inner nature of the pieces doesn't mean that one cannot speak of those rules as of something from which certain consequences flow. For example, it is perfectly correct to say that

(X) One cannot force mate with only a king and a knight.

—is a necessary truth that follows from the rules of chess. Indeed, every chess problem is a challenge to demonstrate a necessary truth (e.g. that in a certain position White can mate in 3 moves) in the conceptual system defined by the rules of chess. Similarly, we can agree with Wittgenstein that the rules of arithmetic should not be seen as derived from some Platonic entities, but that is no reason to deny that those rules have logical consequences that can be drawn out in calculations. Thus, to calculate that $339 \times 275 = 93{,}225$ is to show that this equation follows from the rules of arithmetic.

Or should we not rather say that '$339 \times 275 = 93{,}225$' partly 'constitutes' the meaning of '×'? No, that would be misleading. For unlike the times tables we memorise at primary school and which can justly be said to constitute our concept of multiplication, that particular equation has no pedagogical or normative function. It is not part of the 'rule book' of arithmetic—just as the chess problem in this week's *Sunday Times* is not part of the rules of chess.

And even with respect to the most elementary multiplications, such as '*3 × 3 = 9*', which can appropriately be said to constitute the concept of multiplication as we use them to teach that concept, it is also not incorrect to say that they follow from our semantic conventions. After all, '*p*' follows from '*p. q*'; hence, from the set of rules or grammatical propositions constituting a certain concept we can derive any one of its members.

So far, I see no reason for Wittgenstein to reject the logical empiricist idea that necessary truths follow from semantic conventions. However, mathematical conventionalism has encountered some strong opposition. W.V.O. Quine and Michael Dummett objected that conventionalism is either circular or cannot account for the logical implications of conventions. Crispin Wright tried to show that conventionalism falls foul of an infinite regress. It has also been argued that only an implausibly radical form of conventionalism could withstand the critical implications of Wittgenstein's rule-following considerations. Finally, it can be objected that conventions are arbitrary, which doesn't seem to be true of mathematics. In this chapter, I shall discuss those five objections to conventionalism in turn:

(i) Quine's circularity objection;
(ii) Dummett's objection that conventionalism cannot explain logical inferences;
(iii) Crispin Wright's infinite regress objection;
(iv) The objection to 'moderate conventionalism' from scepticism about rule-following;
(v) The objection from the impossibility of a radically different logic or mathematics.

8.1 Quine's Circularity Objection

To begin with, I shall consider Quine's critical disscussion of conventionalism in his paper 'Truth by Convention' (1936).

Quine proposes to explain the truth of an analytic statement, such as:

(1) A bachelor is an unmarried man.

—as follows: Since the word 'bachelor' is defined to mean 'unmarried man', (1) is equivalent to:

(2) An unmarried man is an unmarried man.

And that is a truth of logic (Quine 1936, 323).—However, that is not a very plausible account of analyticity, as it regards language from the artificial, not to say warped, perspective of formal logic. Logicians may see nothing unnatural in a formula of the form 'A = A', or '$(x)(f(x) \supset f(x))$', but (2) is not at all an ordinary English sentence. Figures of speech apart (e.g. 'War is war'), we have no use for such a reduplication of predicates, it is vacuous or, in the terminology of the *Tractatus*, senseless [*sinnlos*]. Saying that a predicate applies to things to which it applies is comparable to lifting up a chess piece and emphatically putting it down again on the same square. That is not a move in the game, and similarly, one could well imagine that in our natural language we *might* shrug off sentences such as (2) as ungrammatical. Just as we teach our children that a grammatical sentence must have subject and predicate, we might well make it another learners' grammar rule that subject and predicate must be different (cf.: 'a chess piece must be moved to *another* square'). Expressions of the form 'A = A' may of course be used in poetry ('A rose is a rose is a rose') or for rhetorical effects, but just like 'Bachelors, oh, bachelors!' they don't count as declarative sentences, so the question of truth or falsity does not arise.

To be sure, we don't as a matter of fact dismiss sentences such as (2) as ungrammatical, but the fact remains that we don't use them, because, in a natural sense of the word, they don't *say* anything: they are empty and pointless. It is a psychological matter that when forced to call them either 'true' or 'false', we find it more natural to call them 'true', but as far as the actual workings of our language are concerned, we might just as well call them 'nonsense' (cf. *PI* §252). Hence, for Quine to explain the truth or correctness of an ordinary analytic sentence such as (1) as based on the alleged truth of such a linguistic anomaly as (2) is rather perverse.

A much more natural and plausible explanation of the truth of

(1) A bachelor is an unmarried man.

—is to say that it is based on, and an expression of, a semantic norm or convention, namely (Def): that the word 'bachelor' is correctly applied to unmarried men (and nothing else).

It is not an effective objection to the more natural view of analyticity (as due to semantic norms, or the meanings of the words involved) to protest that the existence of such linguistic norms is a contingent matter, whereas analytic truths are supposed to be necessary. This objection rests on a confusion of the internal and the external perspective on a rule-governed activity (cf. Hart 1961, 86–7). The rules of chess, for example, are regarded from an internal point of view as fixed and non-negotiable when one is playing chess. That in a certain game the bishop moved from c1 to f4 is a contingent matter; a different move with the bishop or another piece might have been made instead. But that the bishop was not moved from c1 to c2 is *not* a contingent matter, for such a move is illegal. *Within chess* it is a necessary truth that bishops can move only diagonally, for such are the rules of the game. Again, in a certain position a mate in three moves can be *forced*. Chess problems are based on the *necessity* that is produced by the rules treated as fixed and unchangeable: In response to White's move, Black *must* move and he *can only* move in such and such a way. That is a necessary truth in chess, obviously due to nothing but the rules, which from the internal point of view of chess players are absolutely binding. And yet, of course, there is also an external point of view from which one can describe the origin and development of the game. Here, from a historical or sociological point of view, the same rules are just contingent conventions, which have changed in the past and may change again, should we at some point decide to play a different version of the game of chess instead. Similarly, we can adopt an external perspective on linguistic meanings: their origins and changes over time, but that in no way detracts from their normative force when, taking up an internal perspective, we accept and apply them as they are. While a game is being played and the rules accepted, those rules create necessity, i.e. the *must* and *must not* of valid norms.[2]

Having missed the most plausible construal of analytic truths, Quine suggests that 'truth by convention' cannot be due to definitions, as they are only conventions of notational abbreviation (Quine 1936, 322), available to transform truths, but not to found them. Rather, we must look for another sort of convention, namely postulates (331), 'assigning truth' to a certain kind of statement (334). He then proceeds to set up logic axiomatically, presenting three postulates or axioms that suffice for developing the propositional calculus (one of them corresponding to the inference rule of *modus ponens*: licensing the assignment of truth to any 'q' given the truth of '$p \supset q$' and 'p'), and hinting at four more to cover the predicate calculus as well. Finally, he presents the following problem:

> Each of these conventions is general, announcing the truth of every one of an infinity of statements conforming to a certain description; derivation of the truth of any specific statement from the general convention thus requires a logical inference, and this involves us in an infinite regress.
>
> (Quine 1936, 342)
>
> In a word, the difficulty is that if logic is to proceed *mediately* from conventions, logic is needed for inferring logic from the conventions. Alternatively, the difficulty which appears thus as a self-presupposition of doctrine can be framed as turning upon a self-presupposition of primitives. It is supposed that the *if*-idiom, the *not*-idiom, and so on, mean nothing to us initially, and that we adopt the conventions

2. For a more detailed discussion of analyticity, see Schroeder 2009a.

> (I)–(VII) by way of circumscribing their meaning; and the difficulty is that communication of (I)–(VII) themselves depends upon free use of those very idioms which we are attempting to circumscribe, and can succeed only if we are already conversant with the idioms.
>
> (Quine 1936, 343)

If instead of an axiomatic system we use truth tables to present the propositional calculus (as Wittgenstein did in the *Tractatus*), we can explain the *if*-idiom by the following diagram:

p	q	$p \supset q$
T	T	T
T	F	F
F	T	T
F	F	T

The diagram is to be taken to mean that *if* 'p' is true and 'q' is true then '$p \supset q$' is true, *if* 'p' is true and 'q' is false then '$p \supset q$' is false, etc. Thus, in explaining the *if*-idiom, symbolised by the horseshoe, we already use the *if*-idiom. That is the infinite regress, or circularity, Quine is concerned about: we cannot explain, and thus set up, logic (logical concepts) without already using logic (logical concepts). One could also put it, more generally, like this: Every use of language involves logic, yet one cannot explain logic without language. In short, one cannot explain logic (or language) without already using logic (or language).

The point is a familiar one: one cannot learn one's mother tongue with dictionary and grammar book: through definitions and lists of grammatical rules. Children acquire their first language by imitation and practice, instead. And they are certainly not told during the first stages of learning that the sounds and meanings of our words are conventional. They learn that the colour of grass is called 'green' long before realizing that there are different names for it in other languages and that it could have been different in ours. Does that mean that linguistic meaning and grammar are not in fact conventional? Certainly not.

Towards the end of his article Quine comes close to acknowledging as much, considering that 'it may be held that we can adopt conventions through behaviour, without first announcing them in words; and that we can return and formulate our conventions verbally afterwards, if we choose, when a full language is at our disposal' (Quine 1936, 344). And although he concedes that 'this account accords well with what we actually do' (344), in the end it seems to him too vague and insubstantial:

> We may wonder what one adds to the bare statement that the truths of logic and mathematics are a priori, or to the still barer behavioristic

statement that they are firmly accepted, when he characterizes them as true by convention in such a sense.

(Quine 1936, 344–5)

These questions are not so difficult to answer. That logic and mathematics are a priori means that we can verify their statements without recourse to experience. That is an epistemological observation in need of explanation: How can there be statements—apparently assertions about the way things are—whose truth does not depend on the way things are found to be in the world? A plausible explanation of the apriority of logical and mathematical propositions is that their truth is due solely to the conventional meanings of the words or symbols involved (rather then, say, to some alleged faculty of intuition). To the extent to which we are familiar with their meanings, then, we have no need for further experience in order to convince ourselves of the truth of such propositions.

It is certainly true that an account needs to be given of what it means to say that something is conventional when it was never explicitly introduced as such. The criterion for something being a convention is certainly not (as Quine seems to suggest) that it is 'firmly accepted': Numerous empirical truths have been firmly accepted, without thereby becoming mere conventions, while on the other hand, the acceptance of a convention can be more or less firm: even while a convention is still in force people may be half-hearted about it, regularly considering alternatives.

In order to clarify the concept, consider as a clear example of a convention the use of the English word 'blue' (cf. Hart 1961, 54–6; Schroeder 1998, 41–50):

(i) There is far-reaching *agreement* among our linguistic community about the correct spelling, pronunciation, and application of this word.
(ii) Spelling, pronunciation and use of the word are, however, in a certain sense, *arbitrary*, i.e. not forced upon us by the facts of nature. A different sign with a different pronunciation would be just as serviceable, as illustrated by the equivalent words in other languages. What is more, we are not even compelled by nature to have a word with exactly that meaning. As is well known, the boundaries between different colours are conventional, too, drawn differently in different languages. In Russian, for example, there is no equivalent for the English 'blue', but one word for 'dark blue' [*синий*] and another for 'light blue' [*голубой*].
(iii) The standard spelling, pronunciation and use of the word are consistently kept and conveyed to new members of the community. Deviations are corrected and those corrections are normally accepted. Significantly, such corrections are based only on the fact that a certain linguistic norm is actually in force; it is not required for an

> appropriate correction of a deviation that the norm in question be intrinsically justified. Thus, to be entitled to correct someone's spelling, pronunciation or use of the word 'blue' you only need to point out that it is not in agreement with common usage; you do *not* need to argue that it is a good thing for that English word to be spelt, pronounced or applied as it is. That is a crucial feature of the conventionality of a rule (as opposed to its functionality, for instance): the standard of correctness is constituted by social agreement, and therefore *criticisms of deviation need to refer only to that social agreement or acceptance*, regardless of whether what is thus socially accepted is intrinsically reasonable or better than possible alternatives.

If, according to those three criteria, linguistic meaning is conventional, so is logic. For logic is just an abstraction from linguistic meaning where it concerns the relations between the truth or falsity of sentences (cf. Schroeder 2009a, 87–9). In a broad sense of the term 'logic', it follows logically from the statement that Jones is a bachelor that Jones is unmarried. Taken in that broad sense, all word meanings are relevant to logic. In a narrower sense of the term, logic is concerned with relations between the truth or falsity of statements that depend only on certain structural words, such as 'not', 'and', 'or', 'if', or 'all', together obviously with the terms used to explain the logical features of such connectives, namely: 'true', 'false', 'proposition', 'implies', etc. Either way, insofar as word meaning is conventional, so is logic, which merely reflects certain semantic aspects of our language. Using words with certain meanings *ipso facto* involves using logic. For any substantive change in inferences we draw and accept is a change in meaning (*RFM* 107cd). (For example, if '$p \vee q$' is taken to imply 'p', then '$\vee$' cannot mean the same as our 'or'.)

One may object that since what is conventional could be otherwise, logic cannot be conventional: for after all, one cannot think illogically.—The first reply is that, of course, logic could be otherwise. It is a matter of logic that '$p.\ q$' implies 'p'; but we could easily introduce a rule forbidding this inference. Obviously, such a rule would change the meaning of '$p.\ q$'. 'Logical' means: in accordance with meaning. That one cannot think illogically—that one cannot go against meaning (on pain of producing nonsense), doesn't show that meanings cannot change, and with them our logical inferences.—Against this one may want to say that, of course, words could have different meanings; but given their current meanings, their implications couldn't be different. That, it would appear, is the hardness of the logical must, much firmer than mere convention!—However, the implications just *are* an integral part of the meanings. So what the revised objection boils down to is this: *Holding on to the words' current meanings*, their meanings couldn't be different. And that's not saying anything.

8.2 Dummett's Objection That Conventionalism Cannot Explain Logical Inferences

Michael Dummett, in his influential review of the first edition of Wittgenstein's *Remarks on the Foundations of Mathematics*, presents another, though related, criticism[3] of the logical positivist account of mathematics. 'Modified conventionalism' is Dummett's label for the logical positivist view that only some necessary truths are 'straightforwardly registers of conventions we have laid down; others are more or less remote *consequences* of conventions' (Dummett 1959, 494). Dummett objects:

> This account is entirely superficial and throws away all the advantages of conventionalism, since it leaves unexplained the status of the assertion that certain conventions have certain consequences.
>
> (Dummett 1959, 494)

More recently, Dummett's objection has been urged by Michael Wrigley:

> The more usual form of conventionalism, associated with Logical Positivism, held that certain basic necessary truths owed their necessity purely to our having an explicit convention to that effect, and that all other necessary truths were consequences of these basic conventions. This theory of necessity is immediately attractive because it removes the epistemological mystery from necessary truth. Its crucial flaw, however, is its inability to explain this notion of consequence. The fact that such-and-such basic conventions have such-and-such consequences is a necessary truth but it cannot be a basic convention. What then is the source of its necessity?
>
> (Wrigley 1980, 349–50)

However, it is difficult to see the force of this objection. The picture seems to be something like this: We stipulate a set of axioms, say, the nine axioms of Frege's *Begriffsschrift*:

1. $\vdash A \supset (B \supset A)$
2. $\vdash [A \supset (B \supset C)] \supset [(A \supset B) \supset (A \supset C)]$
3. $\vdash [D \supset (B \supset A)] \supset [B \supset (D \supset A)]$
4. $\vdash (B \supset A) \supset ({\sim}A \supset {\sim}B)$
5. $\vdash {\sim}{\sim}A \supset A$
6. $\vdash A \supset {\sim}{\sim}A$

3. Perhaps Dummett's criticism was inspired by Quine's remark, quoted above, that: 'logic is needed for inferring logic from the conventions'.

7. $\vdash (c = d) \supset (f(c) \supset f(d))$
8. $\vdash c = c$
9. $\vdash (x)\, f(x) \supset f(c)$

How then do we get from those axioms to any other logical truth, not among them, for instance:

(c1) $(x)[f(x) \supset g(x)) \supset (g(a) \vee {\sim} f(a)]$

Presumably, (c1) is a consequence of the axioms, but how it follows from them hasn't been explained.

Something like that would appear to be the picture behind Dummett's complaint,—but it can be quickly dismissed by simply completing the account of Frege's calculus. For those nine axioms are not the only conventions in *Begriffsschrift*. There are also three derivation rules (viz. generalisation, modus ponens, and a substitution rule), which provide a formal explanation of what in this calculus is to count as a 'consequence' of a given formula, and by means of which it is very easy to derive (c1).

That is the obvious answer to Dummett's criticism: Conventions need not take the form of axiomatic statements, they can also be procedural rules, in particular: inference rules, to make explicit the idea of a logical consequence (cf. Bennett 1961). Hence, in the case of an axiomatic system, the idea that a conventionalist, such as Ayer, would lack the resources 'to explain the notion of consequence' is quite groundless.

What about necessary truth in natural languages? As explained above, logical and analytic truths are due to the meanings of words. For example, it characterises the meaning of the word 'if' that a statement of the form '*p*, and if *p* then *q*' implies '*q*'. If we did not acknowledge this consequence we would *ipso facto* have given a different meaning to the word 'if'. To the extent to which the meanings of words have been fixed, the logical consequences of statements made up of those words have been fixed, too. Any unclarity about the logical implications of a statement is an unclarity about the statement's meaning (*RFM* I §§10–11: 41–2, 398b; cf. *BT* 297). So at closer inspection, Dummett's worry is just inconsistent. The idea (to which he thinks conventionalism would be committed) that we might understand a set of explicit linguistic conventions (and hence the vocabulary from which those conventions are formulated and which they partly explain), without yet understanding, or being able to work out, how other things follow from those conventions (and are thus also conventionally determined), doesn't make any sense.

In a word, Dummett's mistake is to think of logic as something on top of meaning. The philosophical picture here is that you can understand the meanings of all the words and statements—and yet not know (nor be able to work out) what logical relations obtain between those statements.

Then of course those logical relations, as something separate from meaning, begin to look rather mysteriously 'unexplained'. It is indeed hard to understand what could be the source of such a free-floating, ethereal mechanism of necessity.

8.3 Crispin Wright's Infinite Regress Objection

Something like, or broadly in agreement with, our response to Quine and Dummett in the preceding sections was already proposed by Jonathan Bennett (1961), trying to show how conventionalism can also explain the notion of logical consequence. Crispin Wright, however, discusses and rejects this reply as unsatisfactory, claiming that it falls foul of an infinite regress, which can be presented as follows:

(1) On the view under discussion ('modified conventionalism'), all necessary truths are *either* conventions *or* their implications.
(2) Suppose a set of conventions *C* implies a statement *Q*.
(3) Now, what is the status of this second-order statement [i] '*C* implies *Q*'?
(4) Expressing a conceptual truth, (i) must be a necessary truth, too.
(5) Hence, according to the view under discussion, it must be *either* an explicit convention *or* an implication of conventions.
(6) As it's not an explicit convention, it must be an implication of conventions.
(7) But the only conventions on which the truth of (i) depends are the set C.[4]
(8) Hence, [ii] '*C* implies "*C* implies *Q*" '.
(9) But (ii) must be a necessary truth, too.
(10) And not being a convention itself, (ii) must be the implication of conventions.
(11) And again, the only relevant conventions are the set *C*.
(12) Hence, [iii] '*C* implies "*C* implies '*C* implies *Q*' " '.

And so on, *ad infinitum* (Wright 1980, 347–50).

Wright then argues that this infinite regress provides a fatal objection to the standard ('modified') conventionalist view:

> The model thus appears to require that in order to recognise the status of any consequence of initial logical consequence conventions, we have to recognise the same of infinitely many statements.
> (Wright 1980, 351)

4. Let us assume that C also contains conventions governing the use of the word 'imply'.

In other words: Wright claims that, according to 'modified conventionalism', in order to understand that *Q* is implied by a set of conventions *C*, we'd have to understand first that '*Q* is implied by *C*' is itself implied by *C*. And in order to understand that, we'd have to understand first that '"*Q* is implied by *C*" is implied by *C*' is implied by *C*, and so on and so forth. Hence, in order to understand any inference from a given set of conventions, we'd have to understand an infinity of inferences—which is impossible.

However, that conclusion does not follow. The infinite regress line (1–12) shows how a derived necessary statement allows the construction of a higher-order necessary statement, for which in turn we can construct a higher-order necessary statement, and so on and so forth. But that does not mean that we *have* to embark on this endless series of constructions. Indeed, it's not even clear that in order to be aware of a given statement's derived necessity we have to be aware of the *possibility* of endlessly constructing meta-statements in this way.

Consider the following analogous argument:

> Suppose *S* is an English sentence. In order to understand the linguistic meaning of *S* we have to recognise that:
> (i) *S* is an English sentence.
> But (i) is itself an English sentence. In order to understand the linguistic meaning of (i) we have to recognise that:
> (ii) (i) is an English sentence.
> And so on indefinitely.

However, in order to understand that *S*—say 'It's raining'—is an English sentence, you *don't* have to recognise that '"It's raining" is an English sentence' is itself an English sentence. There is no need to consider that sentence at all. After all, your understanding of 'It's raining' need not even be formulated in a sentence.

Or again: if a statement *S* is true, then '*S* is true' is itself true. And so is: "'*S* is true" is true', and so forth. You *can* consider, and convince yourself of, the endless possibility of iterating the truth predicate; but you don't have to. You can simply convince yourself that a given statement—'It's raining'—is true (say, by looking out of the window), without considering any such possible iterations.

Like the 'truth'-predicate, the predicate 'is analytic' can always be applied to results of its applications (provided the initial sentence is quoted and not just referred to by a label or an incidental description). '"A bachelor is an unmarried man" is analytic' is itself analytic. Again, we can continue the series, but we don't have to.

Similarly, if we present analytic truths as consequences of a set of all semantic conventions (as in Wright's argument), we can easily convince ourselves that this set entails not only a given analytic truth, but also that statement of entailment itself, and so on and so forth. But so what? It does not follow, as Wright seems to think, that, on the moderate conventionalist view, understanding the initial statement (say, that *S* is analytic) would require that we run through the whole series of iterations: accomplishing 'infinitely many such feats of recognition' (Wright 1980, 351). Indeed, we need not even *consider* the possibility of such endless iterations.

To recapitulate: None of the objections to (moderate) conventionalism considered so far are convincing. Quine is concerned that explicit statements of linguistic conventions presuppose the use of linguistic conventions, but he half admits himself that conventions need not originate with explicit formulations. Dummett complains that moderate conventionalism leaves unexplained how certain conventions can have certain consequences, but with respect to formal systems that is patently mistaken since the concept of a logical consequence is explained by conventional rules of inference, and with respect to ordinary language it is inconsistent since the understanding of logical implication is simply an aspect of the understanding of linguistic meaning: you cannot have the latter without the former. Finally, Wright argues that statements of logical inference imply an infinite series of meta-statements which, *per impossibile*, one would have to recognise in order to understand the initial inference, but, as explained, that is a non sequitur: the possibility of endlessly constructing such meta-statements does not establish the necessity to do so.

8.4 The Objection to 'Moderate Conventionalism' From Scepticism About Rule-Following

However, Dummett and Wright believe that there is yet another, more radical and devastating objection to moderate conventionalism, namely Wittgenstein's rule-following considerations. On their reading (largely in agreement with Kripke's), Wittgenstein has presented a sceptical problem with the very notion of conventional semantic rules. On this view, Wittgenstein argues that it is never 'determined in advance' whether a certain concept applies in a given case, or what is to be the result of a calculation (Wright 1980, 22). That means, (according to this reading of Wittgenstein) it is never determined in advance what is to count as a consequence of a given set of conventions. Therefore, they argue, Wittgenstein could not accept moderate conventionalism, but had to go for 'full-blooded' or 'radical conventionalism', the view that:

> the logical necessity of any statement is always the *direct* expression of a linguistic convention. That a given statement is necessary consists always in our having expressly decided to treat that very statement as unassailable; it cannot rest on our having adopted certain

> other conventions which are found to involve our treating it so. This account is applied alike to deep theorems and to elementary computations.
>
> (Dummett 1959, 495)

Thus for every new calculation or inference, 'we are free to choose to accept or reject' it (Dummett 1959, 495), provided we all agree on our choice: For right is simply what the community accepts (Wright 1980, 226).

It has already been pointed out (in Chapter 7) that as a response to Wittgenstein's rule-following problem the community view is a complete failure. For if it cannot be fixed in advance what in a given case is a correct application of the concept '+ *2*', then it is equally impossible to fix in advance what in a given case is to count as 'community agreement' (*RFM* 392c). Both are on exactly the same footing as instances of Wittgenstein's problem: How can a *general* concept determine its *particular* applications?

For another thing, in an account of mathematics, the distinction between conventions and their implications ('moderate conventionalism'), far from being a weakness or an embarrassment, is definitely what we want (cf. *RFM* 228f; *PR* 248g). For the alternative, the view that all mathematical propositions are conventions is evidently *empirically false*: in conflict with the facts of mathematical practice. As noted above, the mark of conventionality is that the standard of correctness is constituted by social agreement, and that therefore criticisms of deviation need to refer only to that social agreement or acceptance. That is true of basic definitions in arithmetic. How can you justify your insistence that the successor natural number after 6 is 7? Well, that is simply what has been conventionally agreed: what you've been told by teachers and what you find in all the books. But things are very different with the claim that *7957 × 23,249 = 184,992,293*. The reason I can give for insisting on this equation is not that this just is what everybody accepts—I haven't encountered any acceptance of this sum yet, neither by teachers, nor in books, nor by anybody else; rather, my reason for accepting it is that (I convinced myself that) it is what one gets if one carries out a certain kind of procedure for long multiplications. (And note that, *pace* Dummett (1959, 496), there is as a matter of fact no 'putting in the archives' of such new sums either: my accepting that sum *now* will not be invoked by future mathematicians in order to justify *their* acceptance of it afterwards. They will never know of my calculations, and even if they did, the fact that I believe this to be the right result will carry no normative force against the standard procedural criteria.) Empirically speaking, there is no social agreement on this particular sum, there is social agreement only on the general principles of multiplication. Hence it is not the particular sum that is treated as a convention, but the rules of multiplication. It has to be acknowledged, however, that the teaching of general rules involves sample applications. Indeed, at an elementary level 'the expression of a rule is explained by the value, not the value by the rule' (*Z* §301). Our understanding of addition, for example, is based on the simple sums we memorise at primary school, such as: *2 + 2 = 4*, *2 + 3 = 5*, etc. Of those one can indeed metaphorically say that they have been put 'in the archives'. But calculations with larger numbers, which we don't memorise, but have to work out when the need arises, are a different matter. 'Full-blooded conventionalism', refusing to acknowledge the distinction between definitions (and some paradigmatic elementary sums) on the one hand, and calculations (to be done as the need arises), just doesn't agree with our mathematical practice.

Moreover, the idea of 'full-blooded conventionalism' is not only 'hard to swallow' (as Dummett complains), and empirically false, but sheer *nonsense*. Dummett seems to think, like Quine, that conventionality might simply be a matter of firm acceptance of a given statement, treating it as 'unassailable', so that if we decided to hold a given statement as true come what may we would thereby turn it into a convention. Not so.

To begin with, a particular statement is not a convention, however stubbornly one may hold on to its truth. A convention is an agreement what to do (not just what to believe) under certain *repeatable* circumstances, in a certain *kind* of situation, not just on one occasion. Hence a referendum, a one-off decision, is not a convention. Thus we have conventions about how to calculate sums, any sums, i.e. conventions about the use of the addition sign, the multiplication sign, etc. Of course it is conceivable that instead of having the whole system of elementary arithmetic we might only use a few individual formulae, such as '*5 + 7 = 12*'. In other words, the use of the sign '+' might be limited to only a few combinations of numbers. The reason why even in isolation such a single sum could be a convention is that it is general in its application: it is to be used again and again for calculating the overall number of five objects together with seven other objects.

And here already the rule-following considerations come in: For there is a jump from the general formula to its application on a given occasion. (How do I know that '7' doesn't mean '8' on a Sunday? How do I know that wooden objects are to be counted in the same way as metal objects?) Indeed, even the mere reproduction of the same formula is a case of rule-following: Having agreed to the equation *now* does not force me to agree to it tomorrow. (Having agreed to 'Today is Sunday' today, I will reject it tomorrow.) In other words, coming up with a new token of the formula tomorrow because we agreed to one today, is also an inference (cf. *PI* §214).

Generality is essential to the very concept of a convention. Any convention requires applications to countless particular cases, i.e. inferences to what to do on a particular occasion (cf. Wright 1980, 444). Deciding from case to case, as envisaged by Dummett, simply means *not having conventions*. Hence Dummett's idea of conventionality without any inferences is a contradiction in terms.

It is important to note how wide the scope of Wittgenstein's rule-following considerations is. If, like Dummett, Kripke, and Wright, we took them to show that there could be no rule-governed inferences from the general to the particular, we would have to give up on conventions, most notably on linguistic conventions, and hence on general terms in any kind of statement or utterance. On that reading of the rule-following considerations, there could be no concepts with an intension determining their extension. There could simply be no language!

Wittgenstein's famous rule-following argument (of *PI* §§198–201) is a reductio ad absurdum: If you insist on a certain philosophical account of rule-following, rule-following (and hence language) turns out be impossible. So, clearly, that philosophical account of rule-following must be mistaken—'it can be seen that there is a misunderstanding here' (*PI* §201).[5]—Obviously, for such an argument to be understood, the absurdity that is presented as a provisional conclusion must be recognized as such. It is crucial that one sees just how devastating the implications of the view in question are—which is what Dummett and Wright failed to do. Dummett thought that it would undermine just a moderate type of conventionalism, whereas in fact it would do away with conventions, language and all. Wright, like Kripke, thought the damage could be patched up by invoking community agreement, not realising that the recognition of community agreement would itself have become impossible.

As already explained, the philosophical prejudice that is shown by Wittgenstein to lead to absurdity is the view that for it to be determined that a concept *F* applies on a particular occasion *o* it must be *unmistakably and compellingly* laid down somewhere that *F* applies on *o*, and so for every possible application. We are inclined to think that for meaning to be fixed in advance of particular applications, there must be something *in our heads* from which any particular application could be derived with logical necessity.[6] In the first sections of Part 1 of the *Remarks on the Foundations of Mathematics*, Wittgenstein makes it very clear that, of course, meaning *is* determined; only that determination must not be imagined to consist in some sort of infinitely explicit instruction manual in our heads. Meaning is use and cannot be reduced to or be based on mental representation. In short, Wittgenstein's verdict that meaning is not determined *by mental representation, in the head* (see e.g. *RFM* 409c),

5. For a more detailed account of the dialectic of Wittgenstein's rule-following discussion, see Schroeder 2009b, 104–18.
6. Wittgenstein also considers ostensive definitions (*PI* §§27–64) and Platonist ideas of a foundation of meaning (*PI* §§191–7), but most people today, like Kripke and Wright, seem to find the mentalistic line far more tempting.

is turned by Dummett, Kripke, and Wright into the absurdly radical, indeed nonsensical, claim that meaning is not determined *full stop* (e.g. Wright 1980, 22), we can never be committed by any semantic convention (Wright 1980, 232), meanings are always to be freely chosen by us as we go along (Dummett 1959, 495–6).[7]

In conclusion, the rule-following considerations do not provide any convincing objection to 'moderate conventionalism' either. It is an egregious misunderstanding to think that Wittgenstein tried to argue against the possibility of inferences from general statements to particular cases. And if his remarks were taken in such a destructive way they would demolish not only moderate conventionalism, but the very possibility of general concepts, that is, of language.

8.5 The Objection From the Impossibility of a Radically Different Logic or Mathematics

Perhaps the most forceful objection to conventionalism is that whereas what is true by convention (e.g. that 1 m = 100 cm) could easily be changed (we could introduce alternative units of measurement), the truths of mathematics do not seem to be negotiable in that way. The idea that we might have had radically different mathematical conventions whose implications would be incompatible with the truths of our existing mathematics seems entirely unacceptable. It would appear that an arithmetic containing the equation '*7 + 5 = 13*' was simply incorrect. Yet on Wittgenstein's view such a different mathematical system could not be rejected as *false*, although it may well be impractical. Is that a tenable position? Are drastically different rules of arithmetic even conceivable?

Let us first consider logic, of which Wittgenstein appears to take a similar conventionalist view, although there is a significant difference between logic and mathematics: Logic, as remarked earlier, is simply a certain pervasive aspect of our ordinary language, whereas mathematics can be regarded as a fairly distinct extension to our language. All language is governed by logic, while mathematics is more localised: an added specialist language game (roughly speaking, for dealing with quantities).

Wittgenstein repeatedly compares logical inference to a translation from one unit of measurement to another.

> Compare saying that one thing follows from another with changing the unit of measurement. . . . "It has 30 cm, therefore it has so-and-so many inches."
>
> (*LFM* 200–1)

7. Wittgenstein says exactly the opposite: 'When I follow a rule, I do not choose' (*PI* §219).

In fact, at one point he presents transforming the unit of a measurement as an actual example of a logical inference:

> What we call 'logical inference' is a transformation of our expression. For example, the translation of one measure into another. One edge of a ruler is marked in inches, the other in centimetres. I measure the table in inches and go over to centimetres *on the ruler*.
>
> (*RFM* I §9: 40–1)

This is surprising, as the translation is not an a priori one: it does not appear to be based on the meanings of the terms alone (as when I make a calculation based on the definition that 1 inch = 2.54 cm). It is like working out that *17 + 56 = 73* by counting buttons (first, 17, then another 56, then the lot), instead of doing the sum. It wouldn't be an arithmetical calculation, but an experiment (cf. *RFM* I §37: 51–2). Perhaps what Wittgenstein has in mind here is that the double-edged ruler serves as a canonical sample: providing paradigms of both units and of the relation between them (cf. *RFM* 430c). Thus, if both 'inch' and 'cm' are taken to be defined by this particular ruler, the correlation of, say, 12 inches and 30.48 cm read off it, can be said to be derived from the definition of those terms.

In any case, deriving an object's length in cm from its length in inches is certainly an inference of one statement from another. The result is not really a new piece of information, but only a different way of presenting the same piece of information (say, the length of the window). Just as '$\sim(p \,.\, q)$' is the same piece of information as its logical implication '$\sim p \vee \sim q$', only expressed differently. This is the point of Wittgenstein's remark: Logic is not a source of new knowledge, but merely a technique of expressing (parts of) the same claims in different ways.

However, in *RFM* I §5 logical inference seems to be compared not with transformation of the expression of a measurement, but with measuring itself. Inferring according to different rules of logic would be like measuring with different kinds of rulers:

> What would happen if we made a different inference—*how* should we get into conflict with truth?
>
> How should we get into conflict with truth, if our foot rules were made of very soft rubber instead of wood and steel?
>
> (*RFM* I §5: 38)

In what way is measuring like an inference? After all, measurement is not the transformation of an expression, but the production of an expression, a certain description of reality (e.g. 'The book is 7 inches long'), from scratch. Perhaps it could be seen as an inference of the following form:

> The distance between two successive lines on this ruler is 1 inch.
> Laid alongside this ruler the book reaches from the first to the eighth line.
> Therefore, the book is 7 inches long.

Or, for short:

> This book reaches from the first to the eighth line on this ruler.
> Therefore, the book is 7 inches long.

If due to the material of the ruler the distance between the lines is different on different occasions (relative to our customary rulers), we are inclined to dismiss this result as unreliable.

Wittgenstein, however, questions whether we are indeed entitled to find the results of such unconventional measurements likely to be false:

> How should we get into conflict with truth, if our foot rules were made of very soft rubber instead of wood and steel?—"Well, we shouldn't get to know the correct measurement of the table."—You mean: we should not get, or could not be sure of getting, *that* measurement which we get with our rigid rulers. So if you had measured the table with the elastic rulers and said it measured five feet by our usual way of measuring, you would be wrong; but if you say that it measured five feet by your way of measuring, that is correct.—"But surely that isn't measuring at all!"—It is similar to our measuring and capable, in certain circumstances, of fulfilling 'practical purposes'. (A shopkeeper might use it to treat different customers differently.)
> (*RFM* I §5b: 38)

In the cinema, Wittgenstein once saw Eddie Cantor in the film *Strike Me Pink* (1936) play a crafty shopkeeper that uses an elastic yardstick, extending it when measuring out a piece of cloth for a customer, but shrinking it when cutting the cloth (thus delivering less than the customer is expecting to receive) (RR 121–2).

Or again, a ruler may be made of a material that dramatically changes in extension at different temperatures:

> If a ruler expanded to an extraordinary extent when slightly heated, we should say—in normal circumstances—that that made it *unusable*. But we could think of a situation in which this was just what

> was wanted. I am imagining that we perceive the expansion with the naked eye; and we ascribe the same numerical measure of length to bodies in rooms of different temperatures, if they measure the same by the ruler which to the eye is now longer, now shorter.
>
> (*RFM* I §5c: 38)

Such a ruler could be useful if its changes were in sync with the changes of the objects to be measured (RR 121). This is comparable to the way we identify shades of colours by colour samples: The samples look very different in different illumination, but so do the surface colours with which we compare them, so that in spite of their changeability those samples serve us well. Thus, if we imagine having to identify aluminium sticks of the same length at drastically different temperatures, a tape measure with the same thermal expansion would conveniently allow us to ignore the changes in temperature.

Wittgenstein's point is that the measurements with such a changeable ruler should not be regarded as likely to be *false*. They appear (probably) false only if we misunderstand them as an incompetent attempt at our ordinary technique of measuring. As long as we bear in mind that the reading on the flexible ruler is not to be taken as an object's length according to our normal standards, those discrepant results will not be called false, but at worst uninteresting. Each kind of ruler defines a unit of measurement we may or may not find useful. In the latter scenario, the ruler defines a unit of length that is dependent on the temperature, whereas in the shopkeeper's scenario the unit must be understood to be to some extent variably determined by the person measuring.

Crispin Wright, in his discussion of *RFM* I §5, sides with the interlocutor's unsympathetic reaction: 'But surely that isn't measuring at all!' He argues that we have a concept of length as a perceptible property that measurement is to determine more precisely. Hence our measurements, in order to deserve that name, must be roughly in agreement with our observational assessments, whereas the readings of such a soft ruler may differ wildly from our visual impression of an object's length. Moreover:

> It is a feature of the concept of measuring that an accurately measured object will yield distinct readings at distinct times only if *it* changes; so much is implicit in the notion that measuring is to ascertain a property of the object measured.
>
> (Wright 1980, 58)

Yet, the shopkeeper wielding a soft ruler is supposed to produce different 'measurements' for different customers (or at different points during the sales process), knowing full well that the goods themselves have not changed. Indeed, such a shopkeeper's cunning use of a soft ruler would presuppose 'some concept of sameness of length other than is determined by soft-ruler measurement': otherwise he couldn't know that he treated different customers differently. So the soft ruler would not define a new concept of length, but only be an unsuitable means for ascertaining length (Wright 1980, 60).

But is it really true that acceptable measurements of an object must never vary unless the object itself has changed? Frequently in everyday life we only need a rough idea of an object's length, so that measuring procedures that we know to be imprecise (such as using one's stretched out index finger and thumb as a unit) are quite sufficient. But then it's easily conceivable that some unsophisticated people may rely entirely on such makeshift means of measurement. It is also worth remembering that in the past units of measurement were often defined in terms of parts of the human body regardless of differences among individuals. Thus an ell was defined as the length of a man's arm from the elbow to the tip of the middle finger, which would obviously vary considerably from man to man.

As for the correlation between measurement and visual assessment, two things can be said in defence of Wittgenstein's examples: First, a flexible ruler would not yield entirely random results. It can be stretched only to a certain extent, and we can imagine it to be used in a way that when two comparable objects are to be measured one is not supposed to handle it in a very different manner. Secondly, it is quite a common thing for our measurements to contradict and revise our previous visual impressions. So we don't expect a rigorous correlation between the two anyway.

Finally, is the shopkeeper's use of a flexible ruler logically parasitic on a more rigorous concept of measurement? That depends on how we imagine the case. We could of course imagine that the ruler in question could be used in two distinct ways: either like an ordinary tape measure or stretched out to give smaller measurements at will. In that case, it would be more natural to say that the shopkeeper had an ordinary, tolerably accurate ruler—only that it could also be misapplied in a deceitful way by stretching it. But imagine the scenario as follows: The ruler is such that one cannot use it at all without stretching it; in its unstretched state it is all rolled up and so compressed that its markings cannot be distinguished. So even an unbiased use—without any intention of influencing the result in one direction —, depends on the force exerted by the user and so we 'could not be sure of getting *that* measurement which we get with our rigid rulers'.—It is true that where such a ruler is used in an intentionally biased manner the user must be aware of stretching it more or less than he would normally have done: so he must be aware of the distinction between, say, '10 inches' of cloth measured stingily, generously, or without any bias. But it doesn't follow that he must have a 'concept of sameness of length other than is determined by soft-ruler measurement', since even his most fair-minded measurements are carried out with that soft and variable ruler.

Of course, an independent concept of the sameness of length is very likely to arise wherever it is possible to juxtapose different objects measured. Thus two pieces of cloth measured out as equal in length may appear very different in direct comparison, or vice versa. Hence, a more suitable example for Wittgenstein's purposes might be a device for measuring land, or a technique for measuring the length of time (e.g. by reciting a certain sequence of words).[8]

In any case it should be noted that Wittgenstein doesn't altogether contradict his interlocutor's (and Crispin Wright's) outraged objection that 'surely that isn't measuring at all!'. He contents himself with saying that it is at least 'similar to our measuring'—even if it falls short of the more demanding concept of measuring spelt out by Crispin Wright. It is after all, like measuring, a systematic way of answering questions as to objects' length and we can imagine people having such a practice, call it what you will.

Now we must consider the intended analogy between such deviant ways of measuring (or quasi-measuring) and different kinds of logical inference. Granted that a deviant measurement is not false if we understand it to be framed in an altogether different unit to be defined by that practice (not inches, but flexi-inches, so to speak), it is not clear that we can make a similar move with regard to an inference rule allowing the transition:

(L) $p \vee q \quad \Rightarrow \quad p$.

For what is to correspond here to the new unit of measurement? On an ordinary reading (taking the sign '$\vee$' to mean 'or'), this inference is invalid since it is possible for the premise ('$p \vee q$') to be true while the conclusion ('p') is false.

Of course we could simply take (L) as a partial re-definition of the sign '$\vee$', suggesting that it may here be used to mean 'and'. But such trivial, merely notational differences are not what Wittgenstein has in mind when he considers the possibility of different logical or mathematical rules. Rather, the idea seems to be that we have different inference rules: leading us from the same premises to different conclusions (cf. *RFM* I §7: 40).

Wittgenstein's analogy would suggest that an inference from '$p \vee q$' (taken in the ordinary sense) to 'p' may be accepted as valid—if the conclusion is taken in a different sense. But what sense could that be?—Presumably, the conclusion must be taken as a different kind of statement, perhaps one that does not commit itself to the truth of what is said, but

8. Baker & Hacker (2009, 324–7) give the example of traditional Japanese time measurement.

puts it forward only as a likelihood. Given that two possibilities have been envisaged ('$p \vee q$'), it may be deemed licit to derive either *as a reasonable conjecture*. In that sense, '*p*' can be said to 'follow' from '$p \vee q$'.

'But surely', one may want to object, 'that isn't logical inference at all!' And Wittgenstein's reply would be, as in the analogous case of measuring: It is similar to our logical inferences and capable, in certain circumstances, of fulfilling 'practical purposes'. Of course, if we insist that a logical inference must be 'truth-preserving', such an alternative logic would not be acceptable; but we may not so insist. And as long as we know what we're doing, adopting a logic of reasonable conjectures (some of which may well turn out to be false) does not get us into conflict with truth. After all, even writing fiction doesn't get you into conflict with truth, as long as you're not confused about the role and significance of your words.

What about deviant bits of mathematics? Let us consider the following rather odd exchange:

> "You only need to look at the figure
>
> 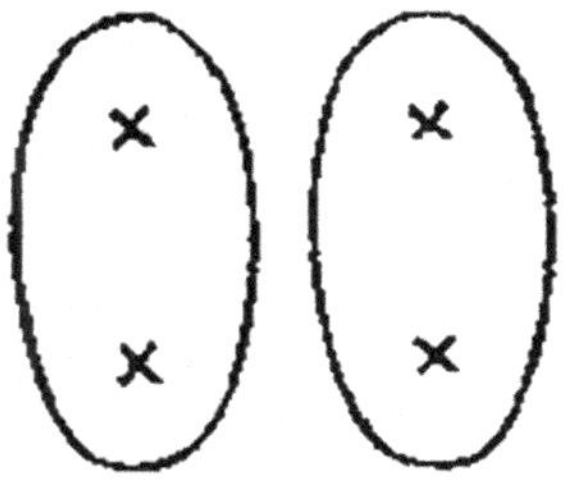
>
> to see that 2 + 2 are 4."—Then I only need to look at the figure
>
> 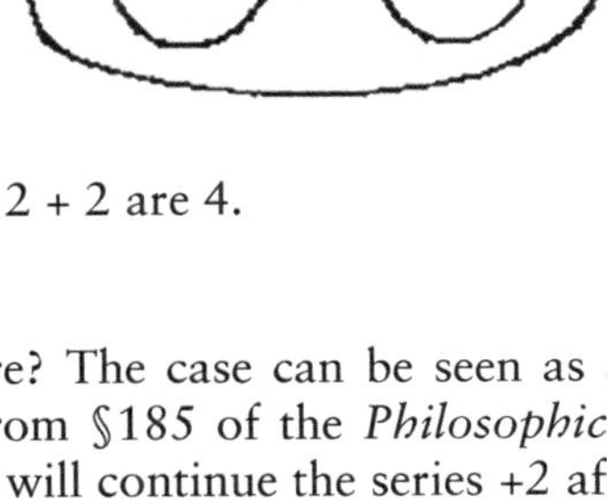
>
> to see that 2 + 2 + 2 are 4.
>
> (*RFM* I §38: 52)

What's going on here? The case can be seen as a variant of the deviant pupil scenario from §185 of the *Philosophical Investigations*. Just as probably all of us will continue the series +2 after 1000: 1002, 1004,

1006;—and not: 1004, 1008, 1012—we will naturally accept the first drawing as a demonstration that 2 + 2 = 4. Or to put it differently: when asked how many pairs there are in 4, we will answer 2 and the first diagram could serve us as a proof. The deviant pupil, however, may claim to have found three pairs in 4, as shown in the second diagram.

Of course we will remonstrate that one mustn't count the same thing twice: as belonging to more than one pair, but the deviant pupil will not be impressed. 'I thought that was how I was meant to do it'. Perhaps he'll argue that the pairs must after all be connected, like the links in a chain.

Wittgenstein's theme, here as in the familiar rule-following scenario, is the idea of compulsion. Of course we accept the former figure as a proof. 'But could I do otherwise? Don't I *have* to accept it?'—Well, we do accept it, quite emphatically, and so we use the emphatic expression 'I have to admit this' (*RFM* I §33: 50).—But are we not *forced* to accept it?—Are we? Not really, is Wittgenstein's reply (*RFM* I §51: 57). After all, this is only a *picture*, and how can a mere picture contain an obligation (*RFM* I §55: 59)?

As in the case of rule-following, we cannot compel people to play along if they stubbornly insist on a deviant understanding. However, whereas in a simple case of rule-following (continuing a series of symbols) that is the end of the matter, in mathematics things would appear to be different. The natural response is that 2 + 2 + 2 = 4 is just false; and if our proof cannot convince the deviant pupil, nature will. For applications of this bizarre equation will constantly result in disappointment. (Buying 4 biscuits will not allow you to give Tom, Dick and Harry a couple each.)

For another example of such an aberrant calculation, Wittgenstein considers the possible results of an application:

> Imagine someone bewitched so that he calculated:

i.e. 4 × 3 + 2 = 10.

> Now he is to apply this calculation. He takes 3 nuts four times over, and then 2 more, and he divides them among ten people and each one gets *one* nut; for he shares them out corresponding to the loops of the calculation, and as often as he gives someone a second nut it disappears.
>
> (*RFM* I §137: 91)

Under such mysterious circumstances—or, in such a strange world, where nuts disappear for no apparent reason—the bizarre equation '$4 \times 3 + 2 = 10$' can be applied successfully.

Suppose—which is of course extremely unlikely—that all of us are bewitched in that way, and that due to the strange behaviour of nuts and other countable things we rarely get into trouble when applying the equation. That is the kind of situation that Wittgenstein envisages in the following remark:

> Imagine the following queer possibility: we have always gone wrong up to now in multiplying 12 × 12. True, it is unintelligible how this can have happened. So everything worked out in this way is wrong!—But what does it matter? It does not matter at all!—And in that case there must be something wrong in our idea of the truth and falsity of arithmetical propositions.
>
> (*RFM* I §135: 90)

In what way do these far-fetched possibilities shed light on the truth or falsity of arithmetical propositions?—They are intended to rebut both the Platonic picture (according to which $12 \times 12 = 144$ is a super hard metaphysical truth) and the empiricist picture (according to which $4 \times 3 + 2 = 10$ is false because nuts and things do not disappear like that).

First, the rule-following considerations undermine the idea of a Platonic realm in which all the steps are already taken. Following a rule, manipulating symbols according to a given recipe, is not spelling out what in some sense is already—and eternally—there. It is essentially a human activity: based on the stability of our attitudes and inclinations. The logical compulsion we experience in applying a rule this way and not that way is man-made: it is our own inexorable insistence on proceeding in one way rather than another (*RFM* I §118c: 82). Socially speaking, it is compulsion in so far as we reject deviant behaviour: we shall not *call* that 'following the rule', 'adding', 'inferring', etc. (*RFM* I §§116, 131, 133: 80, 89, 90). But then, where there is *no* deviancy: where we all agree to apply the rule in a certain way, there cannot be a mistake, for that would require the kind of Platonic standard of correctness Wittgenstein rejects.[9]

9. Note, however, that this does not mean that communal agreement is our criterion of correctness. Rather, beyond appeals to the rule—which in the envisaged cases of deviancy fail to settle the dispute—there *is* no further criterion of correctness. There are only brute facts about what we *take* to be correct. That is exactly one of the conclusions Wittgenstein draws from the case of the deviant pupil—: that we cannot *prove* him to be wrong (cf. *RFM* I §§113, 115: 79, 80). For saying that we *all* do it like that—is not a proof.

Hence, when we imagine the kind of weird behaviour of the deviant pupil to be constantly *our* behaviour—which in that case of course wouldn't appear weird to us—, it can no longer be called erroneous.

So much for rule-following and the lack of Platonic underpinnings. But then, mathematics is more than producing formulae according to the rules of a calculus. Producing lines of patterns according to the rule 'Two strokes, one circle' (perhaps when producing wallpapers) is not mathematics (*LFM* 34). As Wittgenstein puts it:

> I want to say: it is essential to mathematics that its signs are also employed *in mufti*.
>
> It is the use outside mathematics, and so the *meaning* of the signs, that makes the sign-game into mathematics.
>
> (*RFM* 257de)

The prefatory phrase warns us that this may be in need of qualification.[10] It is certainly not true that every mathematical *proposition* or formula has an application outside mathematics, nor is that Wittgenstein's claim. He only requires mathematical *signs*—or concepts (*RFM* 295f)—to have such an external use. What one should say in qualification is that it suffices for a mathematical sign to be definitionally linked, perhaps only indirectly, with signs that occur outside mathematics (*LFM* 33).

As for mathematical propositions, a line *in the middle* of a lengthy mathematical proof is obviously not meant to be applied, but only a step on the way to something that may be applicable. And even the end result of a proof may be intended for further inner-mathematical work, rather than an application in engineering, say. Moreover, even an indirect outer-mathematical usefulness of a formula may only be a vague hope, rather than a clear intention. Even so: even if the link between a given piece of mathematics is typically only indirect and often just a vague future possibility (RR 132), it can be argued that if this practical dimension was removed entirely, it would be less clear that we would still be concerned

10. Cf. Floyd and Mühlhölzer (2020, 42–9), who, it seems to me, exaggerate the tentativeness of this remark. After all, the idea that mathematics is geared towards application goes back to the *Tractatus* (6.211), and then becomes a core idea in Wittgenstein's account of mathematics as grammar, i.e. as grammar of non-mathematical language (MS 127, 236; see Chapter 6; cf. Baker & Hacker 2009, 262–70). In his 1939 lectures Wittgenstein explains that 'mathematical propositions containing a certain symbol are rules for the use of that symbol . . . in non-mathematical statements' (*LFM* 33; cf. 46). And again: 'All the words of mathematics occur *outside* mathematics also. Mathematics gives rules for operating with them' (*LFM* 268). The claim of *RFM* 259de is reiterated, without any note of hesitation, at *RFM* 295f: 'Concepts which occur in "necessary" propositions must also occur and have a meaning in non-necessary ones.'

with mathematics, rather than something more like a chess puzzle: the investigation of logical relations within a mere game.

Moreover, as the examples of *RFM* I make clear, Wittgenstein's concern (at least from 1937 onward) was for the most part with elementary mathematics, with what he once called 'the beginnings of mathematics' (MS 169, 36v [1949]). That is, he aimed at clarifying an aspect of, or a familiar extension to, ordinary language: arithmetic and geometry, as we all learn it at school. And at least of this most common and important part of mathematics it is certainly true that its symbols and rules are essentially geared towards applications outside mathematics.

So, for an assumed mistake in the way we calculate *12 × 12* (or: *4 × 3 + 2 = 10*) not to matter it's not enough that we (all) find the calculation convincing (even when checking it again and again); we also have to find it applicable. And the strange scenario of *RFM* I §137 illustrates how that could be: In a world in which things vanish unaccountably, but regularly, it might be practical to have a rather odd system of arithmetic, or even of counting (cf. *RFM* I §140: 91–2).

But note the difference between Wittgenstein's account and an empiricist one:

The empiricist view of arithmetic is that equations are general claims about the results of adding or taking away physical objects—:

> Put two apples on a bare table, see that no one comes near them and nothing shakes the table; now put another two apples on the table; now count the apples that are there. You have made an experiment; the result of the counting is probably 4.
>
> (*RFM* I §37: 51)

On the empiricist view, this and similar experiences lead us to the general claim that 2 objects added to 2 objects always result in 4 objects, which is expressed by the equation: *2 + 2 = 4*. Thus, an equation is a universal claim about countable objects that would be refuted if we ever encountered an exception: a case of adding 2 things to 2 things resulting in only 3 things, say.

Wittgenstein, however, realised that that is not the way arithmetic relates to experience—:

> If 2 and 2 apples add up to only 3 apples, i.e. if there are 3 apples there after I have put down two and again two, I don't say: "So after all 2 + 2 are not always 4"; but "Somehow one must have gone."
>
> (*RFM* I §157: 97)

That is to say, when on the basis of my experience with nuts and apples etc. I put forward the sum '*2 + 2 = 4*' it is not presented as an empirical generalisation, but as a rule or norm of representation, which means that it is made immune from empirical falsification. We will insist on it even where experience seems to contradict it: in that case we declare our experience as inaccurate. We say that something else must have happened that we didn't see.

However, even though no conflicting experience can *falsify* a mathematical proposition, repeated conflicting experiences can undermine its usefulness. If putting together pairs of apples frequently resulted in more or fewer than 4 apples, we would have to say that our arithmetic was not applicable to apples; as, in fact, it is not applicable to the process of joining drops of water (one and one make one) or mixing measures of different liquids (one quart of alcohol and one quart of water yield only 1.8 quarts of vodka).[11]

> . . . But if the same thing happened with sticks, fingers, lines and most other things, that would be the end of all sums. "But shouldn't we then still have 2 + 2 = 4?"—This sentence would have become unusable.
> (*RFM* I §37: 51–2)

It would not be false; it might still be regarded as a theorem in a certain calculus; but that calculus would, at least for the time being, have no useful application.

And that is the case of the bizarre piece of maths Wittgenstein appeared to prove in *RFM* I §38: *2 + 2 + 2 = 4*. The suggested pictorial proof *might* convince people, so that they adopt it as mathematical proposition. One could not object to that by urging that they would adopt a falsehood. For in their (admittedly odd) calculus it would then be a theorem. Of course, what we naturally take to be the corresponding empirical statements would be false: If you put three pairs of apples or nuts on a table, and take care that nothing is removed or added, you will certainly count a total of 6 apples, not 4. But the falsity of such an *empirical* statement doesn't make '2 + 2 + 2 = 4' false. It only makes it impractical.

Moreover, we could imagine, in the manner of the example in *RFM* I §137, curious circumstances under which countable objects vanish in a way that makes that queer sum usefully applicable. Or, less fancifully, we might hit upon things or chemical substances that add up in just the way suggested by the second 'proof' in *RFM* I §38: so that pairs 'overlap' in one unit ('2 + 2 = 3'). (Thus this calculus might be used for calculating something like the length of a bicycle chain.)[12]

11. Even in these cases, we can still calculate what we put in (e.g. 2 quarts of liquid); it is just that over time those units don't add up according to our rules of addition.
12. We shall return to this example in Chapters 9 and 10.1.

So, mathematics is not forced upon us as true to the facts; it is to be assessed as more or less practical or useful, relative to our circumstances and purposes. A deviant and bizarre piece of mathematics (such as a method of proof that leads to '$4 \times 3 + 2 = 10$') is not false, but only extremely unlikely to be practical. In *RFM* I Wittgenstein imagines that the circumstances are so strange that the equation in question might actually be practical. In a lecture (1937) he imagines the more likely case that it doesn't work (*LFM* 202–3). The example is slightly different, but of the same kind: using a diagram in which groups of three strokes are taken together, but in a way that those triples overlap in one stroke, it is calculated that $9 = 4 \times 3$.

> Suppose people . . . calculated this way when they wanted to distribute sticks. If nine sticks are to be distributed among three people, they start to distribute four to each. Then one can imagine various things happening. They may be greatly astonished when it doesn't work out. Or they may show no signs of astonishment at all. What should we then say? "We cannot understand them."
>
> (*LFM* 203)

If bad mathematics is not false, but impractical, people that hold on to it in spite of its apparent uselessness are difficult to understand. It's not that they got their facts wrong, it's just that they seem so unreasonable. As unreasonable as those whose mathematical calculations are impeccable, but who calculate what we regard as the wrong things: who apply their calculations in an oddly impractical way. This is the case of the firewood merchants Wittgenstein presents in *RFM* I §§149–50: They pile their wood up in heaps of arbitrary, varying height and then sell it at a price proportionate to the area covered by the piles.

> How could I shew them that—as I should say—you don't really buy more wood if you buy a pile covering a bigger area?—I should, for instance, take a pile which was small by their ideas and, by laying the logs around, change it into a 'big' one. This might convince them—but perhaps they would say: "Yes, now it's a lot of wood and costs more"—and that would be the end of the matter.—We should presumably say in this case: they simply do not mean the same by "a lot of wood" and "a little wood" as we do; and they have a quite different system of payment from us.
>
> (*RFM* I §150: 94)

They may just be stupid (*RFM* I §151: 94; *LFM* 202), but then again, they may have different priorities from ours, so that they are not interested in maximising their profits or savings. By 'a lot of wood' they mean wood covering a lot of ground, knowing full well that the same quantity of wood can be made into 'a lot of wood' or 'a little wood'. Perhaps they appreciate the seller's skill of spreading the logs in a way that they cover a lot of ground and are quite happy to pay more when the article is so skilfully arranged (as many of us are happy to pay more for nice packaging or for some affable smiles from the sales staff).

Crispin Wright is as critical of Wittgenstein's discussion of those curious wood sellers as of his example of elastic rulers. His objection is that Wittgenstein fails to make it plausible that the imagined tribe has a different concept of measuring (in one case) or of quantity (in the other) (Wright 1980, 68). He thinks that Wittgenstein faces a dilemma: either these people's purposes in the described practice are like ours (e.g. determining how much wood they need to buy to last the week), or they are not. In the former case, their practices must be criticised as based on the same concepts of length or quantity, but extremely inept. In the latter case, however, in which we cannot criticize them as inept, we also have no good grounds for attributing to them a concept of *length* or *quantity*. So Wittgenstein fails to show that there could be alternative concepts of measuring or quantity (Wright 1980, 71–2; cf. Stroud 1965, 295).

However, Wright's criticism is curiously unfocused: the result of attributing to Wittgenstein an agenda that is alien to his remarks. Does Wittgenstein aim to describe people with an alternative concept of *quantity*? That is not what he says. Rather, we are told that those people use the expressions 'more wood' and 'less wood' altogether differently: *not* to describe the *quantity* of wood, but to denote an aspect of its arrangement. The strange thing is that they take that aspect as crucial for the pricing of the wood, even though they seem to know that it can easily be changed.[13] The point of the story is not to show that there can be different concepts of quantity, but that we can imagine a social life that is not as focussed on the concept of quantity as ours is, not even when it comes to exchanging goods. We take it for granted that the price of wood must be proportional to its quantity. But different forms of exchange are conceivable, for example that:

> each buyer pays the same however much he takes (they have found it possible to live like that). And is there anything to be said against simply giving the wood away?
>
> (*RFM* I §148: 94)

Strange as those people behave, they don't get their facts wrong. They apply their calculations in a curiously selective manner (only to surface,

13. So they need not be as incomprehensibly stupid as Barry Stroud suggests (1965, 295), but he is right to conclude that such people, presented by Wittgenstein to illustrate the possibility of using radically different concepts, may 'not be fully intelligible to us' (298). For our concepts reflect our concerns and interests and our ways of thinking.

not to volume), but clearly that doesn't make their calculations incorrect. And that is the way, I take it, Wittgenstein wants us to regard cases of strangely different forms of mathematics, too: not as erroneous, not as false claims about reality, but as tools that are used in an odd and probably impractical way. In the wood merchants' case, the tools are like ours, but applied in an odd way which strikes us as stupid; in the earlier example of distributing nuts, the tool itself appeared unsuitable (unless strange things happened). But in neither case are the tools, the calculations as such, to be assessed as false.

If a mathematical equation were a statement of fact, it could be true or false independently of people's beliefs. We could then imagine a whole community to believe erroneously that $2 \times 2 = 5$. But can we really?

> But what would *this* mean: "Even though everybody believed that 2×2 were 5, it would still be 4"?—For what would it be like for everybody to believe that?—Well, I could imagine, for instance, that people had a different calculus, or a technique which we wouldn't call "calculating". But would it be *wrong*? (Is a coronation *wrong*? To beings different from ourselves it might look extremely odd.)
>
> (*PPF* §348; *LW I* §934)

The crucial point is that in order to attribute to a community a considered mathematical belief that $2 \times 2 = 5$ we'd have to imagine them as having a corresponding mathematical practice: a calculus and its applications. What they *believe* in mathematics is a reflection of what they *do* in calculating and applying mathematics. Hence *if* we can indeed imagine them having such a practice, the corresponding belief would be true as a matter of course. So our criticism cannot be that their belief is untrue (it isn't if it correctly reflects their practice), it can only be that their practice is utterly impractical or even ludicrous (cf. *RFM* I §§152–3: 95). Moreover, we should probably not regard it as 'calculating' or a form of what we like to call 'mathematics'.

The same applies to the case of traders that do not insist that the price be proportional to the quantity of the goods sold. We can imagine such a practice, although we find it rather incomprehensible, if not insane. But here, as in the $2 \times 2 = 5$ example above, Wittgenstein reminds us of familiar elements in our own culture that to rational outsiders may appear bizarre, such as the ceremony of a coronation (*RFM* I §153: 95). And as he notes that we may find '$2 \times 2 = 5$' too odd to be called 'mathematics', just as we may refuse to call a deviant transition from one sentence to another (e.g. from '$p \vee q$' to 'p') 'logical' or 'inferring' (*RFM* I §116: 80)—similarly, Wittgenstein would obviously agree with what Crispin Wright meant to present as an objection, namely that those wood-sellers' use of 'more' or 'less wood' does not describe something that we should call quantity.

8.6 Conclusion

'Of course', writes Wittgenstein, 'in one sense, mathematics is a body of knowledge, but still it is also an *activity*' (*PPF* §349; *LW I* §935)—embedded in, and partly constituting, a form of life. Hence, to imagine different, alternative forms of elementary mathematics, we should have to imagine different practices, different forms of life in which they could play a role. If we tried to imagine a *radically* different arithmetic we should think either of a strange world (in which objects unaccountably vanish or appear) or of people acting and responding in very peculiar ways. If such was their practice, a calculus expressing the conventions of representation they applied could not be called false. Rather, our criticism could only be to dismiss such a practice as foolish and to dismiss their conventions as too different from ours to be called 'mathematics'.

In order to imagine radically different elementary mathematics, we would have to imagine a different form of life, possibly in a very different physical environment (e.g. one where objects vanished in strange ways). But that means that, as a matter of fact, no such radically different elementary mathematics is an option for us. For in our world objects do *not* vanish in such a way as to make, say, '*7 + 5 = 11*' regularly applicable without getting us into trouble; nor are we willing to adopt a form of life involving an attitude of indifference to the discrepancies and contradictions such a deviant arithmetic would be likely to engender. Rather, for us the application of mathematics tends to be part and parcel of an attitude of accurate observation and attention to detail. *Given what we (want to) use mathematics for, there appears to be no alternative to something like our mathematics.* (That is to say, there can of course be huge differences in the extent and development of elementary mathematics; as indeed there have been if one considers different cultures at different times. But those are differences of varying sophistication in different areas, not downright contradictions. An arithmetic based on the numbers *1, 2, 3, . . . 10, many*, would be of limited use, but it would not lead to any results in conflict with experience.)

If the upshot of Wittgenstein's considerations is, then, that for people with anything like our form of life a radically different kind of mathematics is not an option, it may appear doubtful whether he should be regarded as a conventionalist (cf. Steiner 2000, 334). After all, conventions, as explained earlier, are characterised by an element of arbitrariness: where it is conventional to act in a certain way, it is not to be justified entirely by the nature of things; rather, the correctness of the action is due to an agreement to do things in a certain way, an agreement that could have been different. Thus, in Britain we drive on the left side of the road and for the colour of ripe tomatoes we use the word 'red', but it is easy to imagine different (contradictory) conventions—to drive on the right and call tomatoes '*rouge*'—that would be equally serviceable. However, it may be argued that our mathematics does in fact contain a comparable amount of conventional arbitrariness. First, there are similar notational differences as in other parts of language. Just as the French have a different word for 'red', they have different names for numbers or mathematical terms. Secondly, that we drive on the left is comparable to the way we write down numbers and sums: from left to right. Thirdly, the classificatory arbitrariness of subsuming cochineal under red, but regarding orange as a different colour is perhaps comparable to our having a number system base 10. And as we found the room for conventional arbitrariness in mathematics restricted by our needs and interests, the same can be said about the conventionality of traffic rules and language. A convention that enjoins people to drive on the right side of the road is equally serviceable, but that instead of agreeing to drive on *one* side of the road we should be enjoined to ricochet from one side to the other is not a practical option. Again, the equivalent of a radically different arithmetical calculus (such as the one suggested by Wittgenstein's Venn diagrams, which yields: $2 + 2 + 2 = 4$) might be colour words whose reference oscillates between complementary colours from one day to the next: say, 'red' applies to tomatoes today and to emeralds tomorrow, and 'green', vice versa.

Wittgenstein was certainly not a 'radical conventionalist', but it seems appropriate to call him a 'moderate conventionalist'. His *conventionalism* is limited by what could be called his *instrumentalism*: that is, his emphasis on the applicability of mathematics. Thus for something to be acceptable as a mathematical concept it must be suitable for people with our form of life (our engineering, mercantile, and scientific interests) in a world like ours (where many objects are reliably countable and measurable).

9 Empirical Propositions Hardened Into Rules

The previous section led us to the conclusion that Wittgenstein's conventionalism was moderated by his concern for mathematics' empirical applicability. It may be useful, then, to pay closer attention to those empiricist or instrumentalist elements in Wittgenstein's thinking.

As noted earlier (in Chapter 6), Wittgenstein suggested that mathematical propositions could be seen as originating from empirical propositions which were at some point 'hardened' into rules (*RFM* 324b, 325b, 437e).[1] He seemed to suggest that our arithmetic is based on experiences with counting objects, experiments such as this one:

> Put two apples on a bare table, see that no one comes near them and nothing shakes the table; now put another two apples on the table; now count the apples that are there. You have made an experiment; the result of the counting is probably 4.
>
> (*RFM* I §37: 51)

Because this kind of experiment leads reliably to the same result, we introduce a corresponding rule, which turns, so to speak, a highly probable outcome into a conceptual necessity.

> I believe that it will probably always be so (perhaps experience has taught me this), and that is why I am willing to accept the rule: I will say that a group is of the form *A* [| | | | |] if and only if it can be split up into two groups like *B* [| |] and *C* [| | |].
>
> (*RFM* I §67: 62)

Thus, after first finding the concepts of *2 and 3* and of *5 empirically* correlated, we come to introduce '2 + 3 = 5' as a *mathematical* proposition, that is: a norm of representation. If now the original experiment leads

1. He considers regarding mathematics as a kind of 'frozen or petrified physics' (MS 123, 17v).

to a different result, we shan't accept it: We shall insist that we must have made a mistake or that something strange must have happened to account for this deviation from our norm.

Not only are elementary mathematical propositions said to be based, genetically, upon corresponding empirical propositions, or experiences, they also require that our experiences continue to be, by and large, in agreement with our calculations. Although no individual experience can disprove an arithmetical equation, used as a norm of representation, a regular discrepancy between rule and experience would undermine the rule's usefulness and eventually make us abandon or change it.

> This is how our children learn sums; . . . one makes them put down three beans and then another three beans and then count what is there. If the result at one time were 5, at another 7 (say because, *as we should now say*, one sometimes got added, and one sometimes vanished of itself), then the first thing we said would be that beans were no good for teaching sums. But if the same thing happened with sticks, fingers, lines and most other things, that would be the end of all sums.
>
> "But shouldn't we then still have 2 + 2 = 4?"—This sentence would have become unusable.
>
> (*RFM* I §37: 51–2)

This is a central aspect of Wittgenstein's account of mathematics that is well worth emphasising. An equation, such as '2 + 2 = 4', is not an empirical generalisation, and hence no contrary experience can disprove it. On the other hand, it is not entirely independent of experience either. It is essentially a norm for describing countable things, like beans and sticks, and hence dependent on its suitability for the purpose (*RFM* 357c).

> The rule doesn't express an empirical connection but we make it because there is an empirical connection.
>
> (*LFM* 292; cf. *RFM* 357c)

The rule's usefulness depends on its continued empirical appropriateness. To return to the trigonometric example discussed in Chapter 6, it is not only that we reject (A2)[2] in the light of (M2):[3] that we insist that some of the measurements of the roof must have been inaccurate. It is also that

2. (A2) The pitch of the roof of my lean-to garage is 15° to the horizontal and the roof extends 5.36 metre horizontally from the wall, and one side of the roof is 1.32 metre higher than the other.
3. (M2) If a right-angled triangle has an angle of 15° and the adjacent side is 5.36 metre then the opposite is 1.44 metre.

when, in such a case, we measure or count again with greater care we shall almost certainly find our empirical observations in agreement with the rule. In this case: if the other measurements prove reasonably accurate, we shall find that

> (E) One side of the roof of my lean-to garage, which is 15° to the horizontal and extends 5.36 metre horizontally from the wall, is indeed about 1.44 metre higher than the other side.

That is to say, we take an empirical proposition, such as this, (E), as confirmation of a mathematical proposition, such as (M2); confirmation not of the *truth* of (M2)—for (M2) is a rule, not a generalisation—, but confirmation of its suitability and usefulness in the light of experience.

We can imagine a tribe of people that, having mastered counting, a system of natural numbers, but do not as yet have any technique of calculating, come to adopt several arithmetic statements as quasi-grammatical rules. In the case of such elementary equations as '*2 + 3 = 5*' this will not make much difference as its applications can be verified at a glance anyway, but suppose they frequently have occasion to put together a dozen of a dozen (of bricks, eggs, or coconuts). Having on various occasions counted that the total number is 144, they adopt the handy rule '12 of 12 is 144'.

Or again, suppose a farmer of that tribe wishes to enclose a rectangular area of 100 square feet. Obviously, he wants to spend as little time and material as possible on fencing. He realises by trial and error that the bigger the difference between the lengths of the sides, the longer the perimeter—and the more costly the fencing. A field of 2 by 50 feet would require 104 feet of fencing; 5 by 20 requires only 50 feet of fencing. It's not difficult to hit on the best solution: a square field takes only 40 feet of fencing. We can imagine that, guided by trial and error, this farming community would soon adopt and pass on the rule that of all rectangles with a given area the square has the least perimeter (cf. Kline 1953, 40).

That was indeed roughly the way the Egyptians and Babylonians acquired the first rudiments of mathematics: not through systematic derivation or proof, but merely empirically, by trial and error (Kline 1953, 40; Meschkowski 1984, 41–2). Obviously, such empirically found mathematical rules or techniques can be inaccurate or even false. The Babylonian techniques for determining the surface area of a circle amounted to taking $\pi = 3$. Presumably, the results would be accurate enough for their practical purposes. However, their methods of calculation of the volume of pyramids or truncated cones were, apparently, quite erroneous, even by their own standards of precision (Meschkowski 1984, 41).

Of a tribe using only a few isolated shortcut rules, such as '*12 × 12 = 144*' or 'square rectangle—least perimeter', we would hesitate to say that they have proper mathematics (cf. MS 127, 225). That is because we would expect mathematics to be systematic. Within a fairly small range that could be achieved by memorising (e.g. all multiplications of one-digit numbers), but very soon there would be too much to memorise. Then what is needed is a technique of calculation that allows one flexibly to produce different results under different circumstances, for example: a technique for multiplying *any* numbers. Indeed, Wittgenstein was inclined to think that *calculation*, a systematic technique of producing new results, was essential to mathematics:

> every proposition in mathematics must belong to a calculus of mathematics.
>
> (*BT* 637)

> Mathematics consists of calculi // calculations, not of propositions.
>
> (MS 121, 71v)[4]

> I should like to say: mathematics is a MOTLEY of techniques of proof.
>
> (*RFM* 176c)

So let us consider an early system of calculation. In ancient Ethiopia, the following system of multiplication was developed (also known as Russian Peasant Multiplication). Of the two numbers to be multiplied the first is progressively halved, the second progressively doubled, in columns written underneath the numbers, till the first column reaches 1. Any remainder of halving an uneven number is ignored. Then the lines beginning with even numbers are crossed out. Finally, the remaining numbers in the right (doubling) column are added up. Here is an example:

~~18~~	~~×~~	~~33~~
9		66
~~4~~		~~132~~
~~2~~		~~264~~
1		528
		594

The practitioners of this early multiplication calculus would not have been able to give any mathematical demonstration of why this procedure could be relied upon. In fact, its efficacy was probably widely regarded as magical, for the reason traditionally given for crossing out the lines beginning with an even number was the belief that even numbers are unlucky (Meschkowski 1984, 34).

4. *Die Mathematik besteht aus Kalkülen // Rechnungen nicht aus Sätzen.* (MS 121, 71v (27.12.1938))

For such elementary mathematics, an empiricist account would appear quite accurate. The 'halving/doubling' multiplication technique cannot, on its own, be regarded as providing proofs (contrary to our ordinary long multiplications, which Wittgenstein often calls 'proofs'), for it is far from obvious why that procedure should yield the result of the multiplication. And we can assume (or imagine) that it was not established through proof or any a priori reasoning, but simply through trial and error. Then, to begin with, its status was that of an empirical generalisation, which subsequently was hardened into a rule (*RFM* 324b). Henceforth, where counting the total yields a different result from that of the multiplication, we reject the count as erroneous (or we suppose that the number has changed).

> And this rule then becomes a standard of measurement. The rule doesn't express an empirical connexion but we make it because there is an empirical connexion.
>
> (*LFM* 292)

So far the successful applicability of mathematics would not appear as a problem. That rules *derived from experience* are found to be *in agreement with experience* is not surprising at all. It is simply a case of induction.

Mark Steiner's reading of Wittgenstein emphasises this point: mathematics is to be perceived as based on observations of empirical regularities, hardened into rules. Steiner suggests that there was a silent revolution in Wittgenstein's thinking in 1937, when the 'hardening' of observations of empirical regularities became Wittgenstein's key idea (Steiner 2013, cf. 2009). That doesn't seem to be entirely accurate since Wittgenstein had presented this view already in lectures in 1934 (*AL* 84), but it is probably true to say that in the mid-1930s there was a gradual shift in his thinking: from a pure a priori account to one that paid more attention to mathematics' empirical underpinnings.

Steiner rightly observes that the view of maths based on empirical regularities appears to be in tension with his other view of maths as grammar given his earlier views about the autonomy of grammar (2009, 10). Perhaps the tension can be resolved, for even if a rule was originally derived from experience, once it has been made independent of it, it can be called 'autonomous'; and the truth of a corresponding empirical proposition doesn't entail the mathematical correctness of the rule in the calculus.

However, Wittgenstein's claim in some of the relevant passages in the early 1930s was that grammar could not be *justified* by reference to reality (*LL* 49, 58, *PG* 184), but that seems to be exactly what one does when pointing out the empirical regularity that made it reasonable to adopt the rule in question. And Wittgenstein's use of the word 'arbitrary' would certainly be rather misleading (cf. Hacker 1996, 274). Adopting the rule '*12 × 12 = 144*' was not at all arbitrary; we couldn't have arbitrarily chosen '*12 × 12 = 77*' instead.

However, Steiner does not restrict his empirical-regularity account of mathematics to the carefully limited elementary maths we have been looking at. He includes all of elementary mathematics, with all its techniques of calculations. He says: 'The proofs themselves, . . . together with the theorems, are drawn from experience' (Steiner 2009, 13). That, I shall argue, is implausible.

Let's move then from the Ethiopian halving/doubling multiplication (whose correctness, we assumed, its practitioners learnt only empirically) to our current techniques of calculation, beginning with addition. (Our long multiplications are based on the memorised table of short multiplications, which can be proven by repeated additions.)

(A) 7 + 5 = 12

Is (A) based on experience? Yes and no. Of course, it can be confirmed, and was possibly first suggested, by experiments with apples and other things. But it is also a necessary, a priori truth. Steiner suggests that its necessity amounts simply to its being stipulated to be correct, i.e. given the status of a rule (Steiner 2009, 12).[5]

To begin with, one might object that what we simply adopt as a rule we don't call necessarily true. One wouldn't call it a necessary truth that bishops (in chess) move only diagonally; it's simply a rule of chess. By contrast, one may well call it a necessary truth, within the framework of the game, that one cannot force mate with only a knight. The difference is that this is not itself a rule of chess, but something that *follows logically* from the rules of chess. And this logical entailment makes us speak of necessity. On the other hand, it doesn't sound wrong to call 'A bachelor is an unmarried man' a necessary truth, although there is no inference involved: it simply states the meaning of a word. Perhaps one could say that although the stipulation wasn't necessary, given that something has been stipulated it follows—trivially—that now it is so. So there is a (trivial) logical inference, after all, and hence necessity.

Be that as it may, as a matter of fact a sum such as (A) has not been individually stipulated as a rule. Rather, we have given normative status to the whole calculus of arithmetic and regard any individual sum as merely derived from the calculus, not as directly stipulated. Moreover, the calculus of addition can be seen as a natural conceptual extension developed from the series of natural numbers as used in the practice of transitive counting. To count objects means that each step in a series of conventionally fixed symbols (1, 2, 3, . . .), from a number to the next

5. The same view has been defended by Sorin Bangu (2017).

number, indicates: 'one more', or 'add 1'. In other words, by our practice of counting things, the number series is already implicitly defined in terms of adding one:

0 and 1 = *1*
1 and 1 = *2*
2 and 1 = *3*
3 and 1 = *4*
4 and 1 = *5*
and so on.

I put '*and 1*', rather than '+ *1*', for at this stage (of $_N$English) we do not yet have the general concept of addition. Addition of any two numbers is a new technique, a new concept, albeit one that people are very likely to develop soon after having mastered the idea of consecutive addition of 1 in counting. Once that concept has been introduced, the definition of numbers can be rewritten as sums:

$0 + 1 = 1$
$1 + 1 = 2$
$2 + 1 = 3$
$3 + 1 = 4$
$4 + 1 = 5$
and so on.

And this series of revised definitions of our counting numbers, together with the concept of addition, suffices to generate, by logical entailment (or step-by-step substitutions of definiendum by definiens or vice versa) any sum. Thus,

(A) $7 + 5 = 12$

—is necessary not (or not primarily) in the sense of a semantic stipulation: as part of the rules. First and foremost it is necessary in the sense that it follows logically from our basic semantic (i.e. pre-mathematical) stipulations concerning counting numbers together with the general concept of addition:[6]

$$\begin{aligned} & 7 + 5 \\ = \; & 7 + \underline{4 + 1} \end{aligned}$$

6. Cf. Ayer 1936, 109–15.—As Frege notes with regard to a similar proof given by Leibniz, here we presuppose the associative properties of addition (Frege 1884, §6). But that presupposition can be justified by the consideration that for our concept of a number of things the order in which we count those things is irrelevant. Hence it makes no difference whether when adding up seven objects and one more and one more we begin with the seven or with the two single objects: calculating $(7 + 1) + 1$ or calculating $7 + (1 + 1)$.

$$
\begin{aligned}
&= 7 + \underline{3 + 1} + 1 \\
&= 7 + \underline{2 + 1} + 1 + 1 \\
&= 7 + \underline{1 + 1} + 1 + 1 + 1 \\
&= \underline{8} + 1 + 1 + 1 + 1 \\
&= \underline{9} + 1 + 1 + 1 \\
&= \underline{10} + 1 + 1 \\
&= \underline{11} + 1 \\
&= \underline{12}
\end{aligned}
$$

As the author of the *Tractatus* put it: 'Mathematics is a logical method' (*TLP* 6.2). Mathematical calculation is a form of logical reasoning: developing the implications of what we laid down in the definitions of our concepts. And the basic concepts of arithmetic are constituted by our practice, first, of transitive counting, and then, extending the counting concept of 'one more' to develop a technique of adding any two numbers of things to calculate the total.[7]

But then, the empiricist position that Steiner attributes to Wittgenstein must be mistaken. A sum, such as '*7 + 5 = 12*', is not derived from empirical observation (or if it is, we can easily find a sum involving greater numbers that isn't)[8] and what is more, it does not stand in need of empirical confirmation, having the same a priori status as a tautology. To see it as based on experience would be just as misguided as the idea that 'A bachelor is an unmarried man' was ultimately derived from empirical findings.

Here it is worth noting that the corresponding experiment of first putting 7 apples on a table, then adding another 5 apples, and then counting all the apples on the table (cf. *RFM* I §37: 51–2) would not be a test of the equation '*7 + 5 = 12*'. For that equation only tells us how many apples have been put on the table, not how many there will be a moment later:

> It may seem as if the equation 5 + 7 = 12 entitles us to make statements about the future, namely to predict what number of shillings I shall find if I count the ones I have in each pocket [5 and 7]. But this is not the case. Such a statement about the future is justified by

7. Note that from a Wittgensteinian point of view, the idea of defining natural numbers by Peano's axioms is as misguided as other foundationalist endeavours. Number words are the most obvious examples of mathematical signs that have meaning outside mathematics (cf. *RFM* 257de). Arithmetic adds to their meaning, but is not to redefine them from scratch as if they were names of independently existing objects.
8. Sorin Bangu, developing Wittgenstein's idea of 'an empirical proposition hardened into a rule' (*RFM* 325b), speaks of a 'two-stage genealogical account' of mathematical necessity: first we find that 'an overwhelming majority of people' get a certain result (the 'peak' result of an experiment), and then that result is conventionally stipulated to be correct (Bangu 2017, 166). In a footnote he concedes that with sums of larger numbers we are unlikely to have such 'peak' results; but then we '"harden" by appealing to *proofs*' (2017, 166; n. 50). But in such cases Wittgenstein's metaphor of hardening is just no longer appropriate, for usually where we rely on proofs there is no empirical generalisation (no 'peak' result) to begin with; nothing empirical to be hardened.

> a physical hypothesis which stands outside the calculus. If a shilling suddenly disappeared, or if a new one suddenly came into existence while we were counting, we should not say that experience had disproved the equation 5 + 7 = 12; similarly, we should not say that experience had confirmed the equation.
>
> (*PLP* 51–2)

Similarly, if one invited only bachelors and spinsters to a dinner party but found that some guests at the party were married, one would not have disproved the analytic truth that bachelors and spinsters are unmarried. The observation would only show that either some people turned up uninvited or some of the invitees got married in the meantime.

Of course it is true that under certain circumstances arithmetic may cease to be applicable. Where the number of a group of things changes unaccountably or just too rapidly, what is now 5 + 7 may a moment later *not* be 12. But then you also find that what is now 12 will later not be 12. In other words, it is not just that we cannot *calculate*, it is that we cannot even *quantify*. Where quantities are too unstable, counting or measuring them becomes pointless since the result will be accurate only for one fleeting moment and then cease to apply. So it is not that *our* arithmetic would become inapplicable (as opposed to another one that might fit); rather, in such a situation, we could no longer usefully determine the number of a group of things, be it by calculating or by counting.

Although it is conceivable that the equation '*7 + 5 = 12*' was drawn from experiences with countable objects—from repeated observations that where we count first seven and then five we can also count twelve—, that is not what Steiner seems to have in mind when he claims that even mathematical *proofs* are drawn from experience (2009, 13). There the relevant experience cited is that of people's regular and predictable behaviour in carrying out calculations. People normally get 625 as the result of multiplying 25 by 25 (*RFM* 325b). It is true that Wittgenstein often stresses the importance of *behavioural* regularities as a transcendental condition of the applicability of mathematical concepts. We could not have a system of arithmetic if people could not be taught reliably to calculate in the same manner. But, clearly, the regularity with which suitably trained people work a given calculus cannot be the empirical data leading to the formation of that very calculus. The behavioural regularities of rule-following are a prerequisite for *any* calculus—also one in which 25 × 25 yields 3. Therefore, such behavioural regularities cannot be invoked to explain why we have *our* arithmetic, rather than a different one: why we use a calculus in which, for example, '*25 × 25 = 625*' is a theorem.

Synthetic A Priori

A sum, such as '*7 + 5 = 12*', is *not* normally derived from experience (as an observation 'hardening' into a rule), but calculated, i.e. logically derived from certain definitions. And yet there remains an interesting difference between an analytic truth, such as 'A bachelor is an unmarried man', and '*7 + 5 = 12*'. You cannot understand both the expressions 'bachelor' and 'unmarried man' without knowing that they are synonymous, whereas (as Kant famously observed) you can well understand the mathematical expressions on each side of a correct equation without yet knowing that they amount to the same number. The predicates 'bachelor' and 'unmarried man' have exactly the same criteria of application; the expressions '7 + *5*' and '12' do not: the first requires you to count first to 7 and then to 5, the second requires you to count to 12. Someone understanding the former, may not even be able to count to 12.

Kant took this difference to show that '*7 + 5 = 12*' is not analytic, but synthetic a priori. Interestingly, Wittgenstein, too, finds it appropriate to call mathematical truths 'synthetic a priori'. His remarks suggest the following reasons:

(i) Unlike tautologies, mathematical equations say something (*WVC* 106). They are informative, may even be surprising (cf. MS 164, 38ff.). '*25 × 25 = 625*' tells me, for example, that 25 boxes of 25 muffins will not suffice to serve one to each of 650 delegates (MS 123, 44r).
(ii) Mathematical propositions need to be worked out, they require a construction, not just a spelling out of implications. Thus, '101 is prime' does not follow from an analysis of the concept of a prime number (*RFM* 247a).
(iii) The result of such calculations is treated as a rule (*RFM* 340c). Thus, our grammar forges the identity between the two sides of an equation, e.g. '25 × 25' and '625' (*RFM* 358–9).

However, it is not clear that these are convincing reasons to regard mathematical truth as synthetic a priori. A contrary view has been persuasively presented by Frege, arguing in his *Foundations of Arithmetic* for the claim that arithmetic is altogether analytic:

> KANT obviously . . . underestimated the value of analytic judgements . . . the more fruitful type of definition is a matter of drawing boundary lines that were not previously given at all. . . . here, we are not simply taking out of the box again what we have just put into it. The conclusions we draw from it extend our knowledge, and ought therefore, on KANT's view, to be regarded as synthetic; and yet they can be proved by purely logical means, and are thus analytic. . . . Often we need several definitions for the proof of some proposition, which consequently is not contained in any one of them alone, yet does follow purely logically from all of them together.
>
> (Frege 1884, §88)

Frege effectively answers the first two of Wittgenstein's concerns. '*7 + 5 = 12*' is indeed not like 'A bachelor is an unmarried man', but it may be compared to the following logically true statement (from Lewis Carroll's logic puzzles):

> Since no interesting poems are unpopular among people of real taste and no modern poetry is free from affectation, while all your poems are on the subject of soap-bubbles, yet no affected poetry is popular among people of real taste and no ancient poem is on the subject of soap-bubbles, it follows that your poetry is not interesting.

This, too, may require some working out, the concluding clause cannot be derived from the analysis of any one concept, and therefore the statement can be informative. One may be aware of a set of premises and yet be surprised by something they jointly entail.[9]

As for (iii), it at least suggests what we rejected as false in the case of a developed mathematical system, namely that the identity between the two sides of an equation is explicitly stipulated in individual cases.[10] In fact, given our number system, we do not *forge* the identity between *25 × 25* and *625*; we *calculate* that there is such an identity.

***

To recapitulate: As mentioned in Chapter 5, Wittgenstein's philosophy of mathematics can be characterized by *two* key ideas, only one of which (that mathematics is grammar) we have been discussing so far. We found, however, that the idea that mathematical propositions are like grammatical norms, owing their exalted status to our endorsement, needed some qualification. For (apart from the fact that mathematical norms do not for the most part belong to ordinary language) such conventionalism is severely restricted by the need for mathematics to be applicable by

9. This is also part of a response to Poincaré's objection against logicism: that it could not explain how mathematics can be an inventive and constantly growing discipline (1905, Ch. 1; cf. Shanker 1987, 282). Moreover, to regard mathematics as analytic is not a commitment to any logicist reductionism. Clearly, the developments of mathematics are partly due the introduction of new concepts, which need not be reducible to the logician's toolkit.
10. And the parenthesis at the end of *RFM* 359a does indeed appear to suggest Steiner's view that arithmetic equations are adopted on empirical grounds. In our arithmetic, sums are not individually given normative status, let alone on the grounds of empirical confirmation. What can perhaps be regarded as forced upon us by experience is numbers: numerous specimens of types of objects and repeated occurrences of certain experiences make it overwhelmingly natural for people to develop concepts for counting. But with an open-ended series of numbers in place, and some equally natural ideas such as adding, removing, or distributing amounts, the development of the rest of arithmetic follows, for the most part, logically.

creatures with certain interests in a certain environment. That is to say, for arithmetical equations to be viable they cannot just be made *(i) true by convention*; essentially they, or to be more precise: their applications, have to be *(ii) empirically true*, too. Yet now we (re-)encounter what seems to be a third source of mathematical truth: mathematical propositions have to be *(iii) true according to the rules of calculation or proof*. How can these three sources of mathematical truth be combined?

(i) Truth by convention
(ii) Empirical truth
(iii) Truth according to the rules of calculation or proof

We saw how the first two could be combined: how empirical findings can 'harden into rules'. A rule may first be suggested by empirical findings and its successful application depends on its continued agreement with empirical findings, yet what makes it (akin to) a convention, rather than an empirical generalisation, is that we will not allow it to be refuted by a few observations to the contrary.[11] But now we have found that the idea that mathematical propositions are derived from corresponding empirical propositions then hardened into rules fits only the unsystematic rudiments of mathematics. Once we encounter a fully developed calculus, such as our school arithmetic, its sums or propositions are logically derived from some basic concepts, rather than inductively derived from corresponding empirical claims.

From an exegetical point of view, it should also be noted that although Wittgenstein occasionally used the picture of observations of regularities being 'hardened into rules', the thrust of his considerations of a mathematical proof goes in exactly the opposite direction. The distinctive feature of a proof is that it does *not* require observations of regularities: that it can convince us of something once and for all, dispensing of the need of empirical confirmation. And Wittgenstein's key idea about mathematical proof (which will be discussed in the next chapter) is that what makes empirical confirmation dispensable is conceptual change.

The same contrast between rudimentary (empiricist) and systematic (logical) mathematics can be put in historical terms. The *empirically based grammar view* of mathematics can be neatly illustrated by *Babylonian* or *Egyptian mathematics*, which consisted largely of empirical observations, measurements, turned into practical rules for the use of engineers and merchants. The *calculus view*, on the other hand,—at least if we take 'calculation' in the broad sense to comprehend any systematic way

11. Though, arguably, the same is true of any well-established high-level empirical generalisation. What makes us doubt the conflicting evidence rather than such a general claim is that the latter is based on a lot more supporting evidence. This has nothing to do with conventionality.

of working out or proving new results (including geometrical proofs)—provides a good description of the achievements of *Greek mathematics*: the studies of a class of gentlemen scholars, not so much motivated by practical needs, which they left in the care of slaves, but fascinated by the idea of timeless and necessary truth as established by a priori proofs.

Let us cast at least a cursory glance at more recent developments in mathematics: at Newton's development of the calculus. Seventeenth-century physics was concerned not only with a body's average speed during a certain amount of time, but also with acceleration. Kepler's Second Law, for example, describes the continuous change of speed of a planet travelling in an elliptical orbit around the sun. Ballistics required the calculation of the velocity—and hence impact—of a cannon ball at a particular point in time. Scientists tried to calculate the velocity of an object during a period h starting from a point in time t. The smaller one takes h to be, the closer the average velocity during h becomes to the momentary velocity at t. In order to calculate the momentary velocity at t, one takes $h = 0$. In physics this method of calculation proved successful, although in mathematics it remained controversial for quite a while. 'The objection to this process is that one starts with an h which is not zero and performs operations such as dividing numerator and denominator by h which are correct only when h is not zero'. Yet, in the end, in order to obtain the *instantaneous* rate of change, h is taken to be zero (Kline 1980, 130).

Calculus was regarded as doubly dubious. It was conceptually problematic in introducing the apparently inconsistent concept of instantaneous change—such as speed without movement. As Voltaire put it, it was 'the art of numbering and measuring exactly a Thing whose existence cannot be conceived' (Kline 1953, 266–7). Moreover, according to existing mathematical concepts, such calculations were regarded as incorrect even by leading mathematicians, such as Lagrange and Euler (Kline 1953, 267). It could not be justified in strictly mathematical terms, but only pragmatically: by its eminent usefulness in modern physics.

This would appear to be a plausible instance of what one could call empiricist conventionalism. An important new mathematical tool cannot be derived logically from existing mathematical concepts and rules. Rather, those rules have to be bent or modified (allowing something like division by zero) in order to produce the kind of formulae required by physicists. It is the empirical success of the calculus, rather than its intrinsic mathematical plausibility that leads to its acceptance in the 17th and 18th centuries. It is probable that more such examples could be found in the history of mathematics: of innovations whose mathematical credentials were, at least to begin with, problematic, but which were accepted, at least partly, on the grounds of their successful applicability. It may not exactly be a matter of empirical propositions 'hardening into rules'; empirical confirmation may be more indirect than suggested by Wittgenstein's metaphor. Still, it would be a matter of mathematical conventions being adopted at least partly on the grounds of empirical considerations.

However, my impression is that those are rather exceptional cases. I'm inclined to believe that normally the development of new mathematical techniques, although perhaps motivated by potential empirical applicability, would have been sound from an intrinsic mathematical point of view, too: rigorously derived from existing concepts and rules.

Be that as it may, even if on the whole, and certainly for our school mathematics, the kind of empiricist conventionalism that Wittgenstein's metaphor of hardened empirical propositions suggests doesn't appear to provide the right model, that does not detract from the plausibility of Wittgenstein's basic idea: viz., that the quasi-grammatical norms of mathematics must be in tune with experience. Where the metaphor of hardening empirical propositions goes wrong is in locating that process of attuning only at the level of *propositions*, when in fact it occurs already at the level of choosing suitable mathematical *concepts*.

As argued above, our arithmetic is developed from our practice of transitive counting, in which the series of numbers is determined by the successor relation of 'one more', i.e. addition of one. In that way, a rudimentary idea of addition is already contained in our practice of counting things. Of course, instead we could have introduced an alternative concept of addition, for example, one in which the terms to be added are taken to overlap in one, resulting in the following sums:

$1 \dagger 1 = 1$
$2 \dagger 1 = 2$
$2 \dagger 2 = 3$
$3 \dagger 2 = 4$
$3 \dagger 3 = 5$
and so forth.

Yet this operation † (let us call it 'overlap addition')[12] does not follow naturally from our practice of counting objects (by continuously adding one). Addition is, like counting, a technique to ascertain the overall number of objects in two sets; so it is closely related to the concept of a number. Overlap addition, by contrast, is not a procedure for ascertaining the total number of objects in two sets. It is an entirely new technique whose result introduces an entirely new concept (an instance of which would be the length of a row of things made up of shorter rows of things overlapping where they are joined up).

And even if instead of our addition we had introduced overlap addition, an arithmetic would have resulted systematically from our concepts and not piecemeal on the basis of empirical observations of *particular* results of joining up rows of things.

12. Inspired by *RFM* I §38: 52, and already considered in Chapter 8.5.

We could also imagine an arithmetic on the basis of a different practice of transitive counting. Suppose we were to count:

1, 2, 3, and, 4, 5, 6, and, 7, 8, 9, and . . .

That is, after every three numbers we inserted the word '**and**' which was also to be correlated with one of the objects to be counted, but not to increase the result. So we count **3** musketeers, but also **3** seasons (**1, 2, 3, and**—result: **3**). Addition would look like this:

2 + 1 = 3
2 + 2 = 3
2 + 3 = 4 or 5
2 + 4 = 6
2 + 5 = 6
2 + 6 = 7 or 8

Thus, it is at least misleading for Stuart Shanker to say that 'mathematical propositions are "rules of grammar" which we *freely construct*' (1986, 21). Free construction happens—here as in other areas—on the level of concepts, rather than propositions. *Concepts* (such as those of our natural numbers) are indeed created in the way Shanker suggests: guided by considerations of usefulness and aesthetics (1986, 23). That kind of 'free construction' is not arbitrary in the sense that radically different concepts would have suited us just as well. It may also be true that with our biological capacities and disposition we could never have settled for dramatically different number concepts (cf. Dehaene 2011). By 'free construction' we only mean that our—or alternative—concepts cannot be regarded as correct or true. Mathematical *propositions*, on the other hand, can be regarded as true: not freely constructed, but derived or calculated from our definitions of mathematical concepts.

It has often been observed that the construction of proofs in higher mathematics requires creative thinking. But as Marcus du Sautoy explained, that is not incompatible with seeing them as developing logical implications. Most of the theorems that could be derived from an existing body of mathematics are 'banal or without interest':

> There is more to mathematics than just generating mathematical truths. The art in being a mathematician is to single out those logical pathways that have something special about them.
>
> (du Sautoy 2011, 21–2)

The mathematician's creativity lies in discovering logical implications that will prove fruitful, both inside mathematics and ultimately in its possible applications.

10 Mathematical Proof

As already said, the calculus view was central to Wittgenstein's thinking about mathematics up to about 1937, when he shifted emphasis to the grammar view, without however retracting the calculus view, still present in his discussion of mathematical proof.

The calculus view is forcefully presented in *The Big Typescript* (1933):

> every proposition in mathematics must belong to a calculus of mathematics.
>
> (*BT* 637)

> Mathematics consists entirely of calculations.
>
> In mathematics *everything* is algorithm, *nothing* meaning.
>
> (*BT* 749)

> For mathematics is a calculus.
>
> (*BT* 750)

As an empirical observation about our mathematical practices this appears uncontroversial, at least if by 'calculation' we understand any kind of technique of proving new results, including geometrical constructions. Not only do we prove or calculate in mathematics, but it seems plausible to say that it is essential to our concept of mathematics that it involves proofs or calculations. In other words, if (as imagined earlier) the equation '*12 × 12 = 144*' occurred in isolation, as a practical maxim, rather than the result of a calculation, we should hesitate to call it mathematics. It may, for those people, acquire the status of a quasi-grammatical rule about quantities and in that respect be very much like our mathematical equations. But there is a crucial difference: it is not the result of a calculation or proof according to certain rules. Yet it is an essential feature of mathematics that its sentences are introduced and justified in a certain way. This is part of their grammar or use (just as it is part of the grammar or use of religious statements that they are not introduced and treated like scientific hypotheses). Mathematical sentences, if they are not definitions giving meaning to new symbols, are neither verified inductively, by repeated tests, nor adopted on authority. They have to be deduced by proof or calculation. That is, they are quasi-grammatical rules: produced by applying second-order quasi-grammatical rules (*RFM* 228f).—Hence, without an arithmetical calculus, the equation '*12 × 12 = 144*' would not really be a mathematical proposition.

The calculus view of mathematics is incompatible with the empiricist version of the grammar view (as presented by Mark Steiner). Mathematical *propositions* (as opposed to concepts) are typically calculated, rather than derived from empirical origins. However, that does not mean that the calculus view is incompatible with the grammar view itself: the normative status of mathematical propositions, as akin to grammatical statements, can be upheld independently of whether they are thought to have empirical or logical origins. The picture that emerges, then, would be that, according to Wittgenstein, mathematical propositions have (as noted before) two essential features:

First, mathematical propositions are given the status of quasi-grammatical rules, norms of representation, constrained by the requirement of applicability (though not, for the most part, based on empirical observations).

Secondly, definitions (and axioms) apart, mathematical propositions have to be legitimised by proof.

Note that the idea of a practice with second-order rules that are not stipulated from the beginning, but can be, or have to be, introduced later according to certain rule-governed procedures is not otherwise unheard of. A common example is a legal system, which consists not only of a set of laws, but also of procedural rules for legally introducing new laws.

Wittgenstein's discussion of mathematical proof centres around three main issues, which we shall consider in turn:

1. What is a mathematical proof? (What distinguishes it from a mere experiment?)
2. What is the relation between a mathematical proposition and its proof? (Is the sense of a mathematical proposition determined by its proof?)
3. What is the relation between a mathematical proposition's proof and its application?

10.1 What Is a Mathematical Proof?

Wittgenstein approaches the question of the nature of mathematical proof by a method familiar from the first pages of his *Philosophical Investigations*, namely by considering artificially simplified scenarios, primitive 'language games', because it 'disperses the fog to study the phenomena of language in primitive kinds of application in which one can command a clear view of the aim and functioning of the words' (*PI* §5). Thus he presents the following primitive example of a mathematical proof:

> But how about when I ascertain that this pattern of lines:
>
>  (a)
>
> is like-numbered with this pattern of angles:
>
> 
>
> (b)
>
> (I have made the patterns memorable on purpose) by correlating them:
>
> 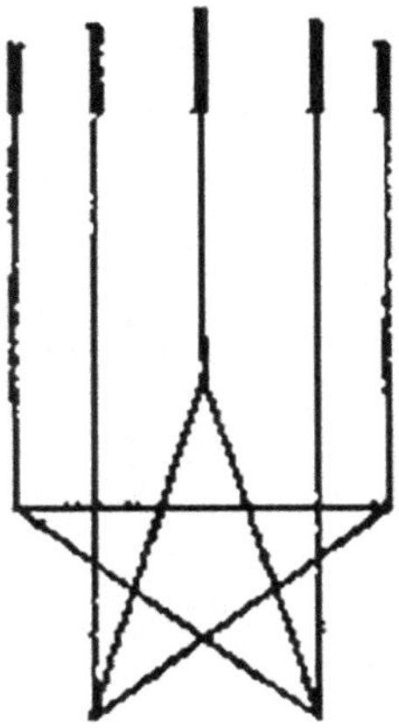
>
> (c)
>
> Now what do I ascertain when I look at this figure? What I see is a star with threadlike appendages.—

But I can make use of the figure like this: five people stand arranged in a pentagon; against the wall are wands, like the strokes in (a); I look at the figure (c) and say: "I can give each of the people a wand."

I could regard figure (c) as a schematic *picture* of my giving the five men a wand each.

> . . .
>
> I can . . . conceive figure (c) as a mathematical proof. Let us give names to the shapes of the patterns (a) and (b): let (a) be called a "hand", *H*, and (b) a "pentacle", *P*. I have proved that *H* has as many strokes as *P* has angles. And this proposition is once more non-temporal.
>
> (*RFM* I §§25–7: 46–8)

The following three important features of a mathematical proof emerge:

(i) What a proof establishes can then be *applied as a shortcut*. At least in the case of elementary mathematics, this application will be outside mathematics. Thus, we can identify groups of wands and people, for example, as instantiating the patterns *H* and *P*. Then this proof of a one-to-one correlation between the respective elements of *H* and *P* allows us to say that an *H*-group of wands is like-numbered with a *P*-group of people. Generally, in future whenever I encounter patterns *H* and *P* I don't have to correlate or count, I can see immediately that they are like-numbered (*RFM* I §30: 49). The proof provides us with 'a new way of establishing numerical identity' (*LFM* 72). That presupposes of course that instances of *H* and *P* are likely to occur repeatedly and that they are straightforwardly recognizable (*RFM* I §27, 41: 47–8, 54).—Similarly, the long multiplication '*21 × 36 = 756*' constitutes a proof of that equation only because now we can apply it to countable objects (*LFM* 36–8). It enables us to tell how many nuts there are in 21 bags of 36 each, without having to count the lot.

(ii) A proof must be *surveyable* (*RFM* I §154: 95). This does not mean that it must be possible to take it all in at a glance (which would obviously be true only of the most elementary proofs), but that it is effectively *reproducible* (*RFM* 143; Mühlhölzer 2006; Büttner 2016a). 'For instance, if [in a long multiplication] the figure on the bottom line were constantly changing, it would be useless to us as a proof' (*LFM* 37).

(iii) Most importantly and provocatively, Wittgenstein maintains that a proof involves an element of *stipulation*: the introduction of a *new or changed concept* (*RFM* 166a-d, 172b, 196f, 237e, 240c, 244d, 297f, 411b, 432f, 434c); what is found in one instance is then accepted—adopted as a rule (*RFM* I §§33, 67: 50, 62)—to hold *generally*. This most original and controversial aspect of Wittgenstein's account of mathematical proof shall be explored in the remainder of this section.

In the *H-P* proof, it is only one particular drawing of a Hand that is found to have as many strokes as a particular drawing of a Pentacle has angles: 'the counting or correlation merely yielded the result that these two groups before me were . . . the same in number' (*RFM* I §31: 49). So how do we know that this will hold for *all* Hands and Pentacles?

> "Why, because it is of the *essence* of *H* and *P* to be the same in number."—But how can you have brought *that* out by correlating them?
>
> (*RFM* I §31: 49)

Wittgenstein has no time for Husserl's idea of *Wesensschau*—the 'intuition of essences'. That is a kind of platonic mythology. Essences cannot be explored or found, they can only be *created* by conventions:

> I deposit what belongs to the essence among the paradigms of language.
>
> The mathematician creates *essences*.
>
> (*RFM* I §32: 50)

Seeing that this Hand and this Pentacle are like-numbered, I feel induced to make their equinumerosity a rule for all Hands and Pentacles.

The same thought is expressed by Wittgenstein's saying that a proof is not an experiment, but the *picture of an experiment* (*RFM* I §36: 51; 160c). The initial drawing of the connecting lines can be regarded as an experiment; but then (when I regard it as a proof) I decide to treat what looks like an experiment as a paradigm by which future instances are to be judged. 'The experimental character disappears when one looks at the process simply as a memorable picture' (*RFM* I §80: 68).

As a proof involves a stipulation, it produces a *change of meaning*. The *H-P* proof changes the meaning of the expression 'having the same number', since it introduces a new criterion for sameness of number (*LFM* 73).—

Wittgenstein anticipates the most natural objection to his account of proof:

> "But now, if he has an *H* of things and a *P* of things, and he actually correlates them, it surely isn't *possible* for him to get any result but that they are the same in number.—And that it is not possible can surely be seen from the proof."
>
> (*RFM* I §31: 49)

In other words, one is inclined to protest that there is no need and no room for any stipulation, since the concepts of Hand and Pentacle suffice to imply the possibility of one-to-one correlation. We seem to be able to perceive, in the drawing that provides the proof, the necessity of their equinumerosity. Yet Wittgenstein insists that no such necessity can be perceived.[1] It *is* possible to find that future attempts at correlation will yield a different result. We might in future omit to draw one of the correlating lines (*RFM* I §31: 49), or we might 'find that (as we should then say) we had always been drawing the lines wrong. We might have always drawn two lines in the same point by a slip' (*LFM* 72)—discovering which we might then conclude that in truth there are one more angles in a Pentacle than strokes in a Hand.

Another primitive example is the proof that four right-angled triangles can be arranged to form a rectangle, like this:

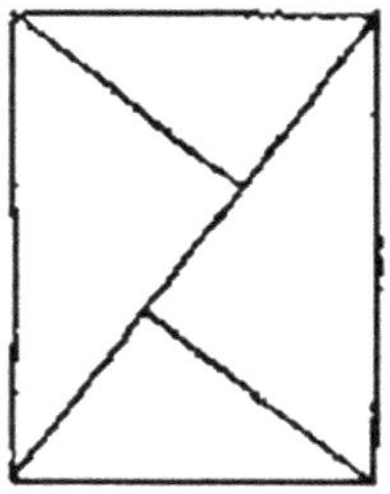

Again, strictly speaking, what we see is only that *these* particular triangles can be arranged to form a rectangle, not the corresponding general proposition about *any* four triangles of those sizes (*RFM* I §42: 55). Since it is easy to imagine that people might struggle in vain to solve this puzzle, how can we now be so confident that *it can always be done*, rather than merely that it was possible on this one occasion (*RFM* I §47: 56)? After all, when Usain Bolt ran the 100 metres in under 9.6 seconds that wasn't a proof that any man could do so.

Wittgenstein's most bewildering example, however, is the one already discussed in Chapter 8:

"You only need to look at the figure

1. In this there is an interesting parallel between Hume and Wittgenstein. Both looked for necessity and couldn't find it. Hume concluded that (in explaining causation) we have to contend ourselves with constant conjunction, whereas according to Wittgenstein necessity is not to be denied, only it is not *found*, but stipulated.

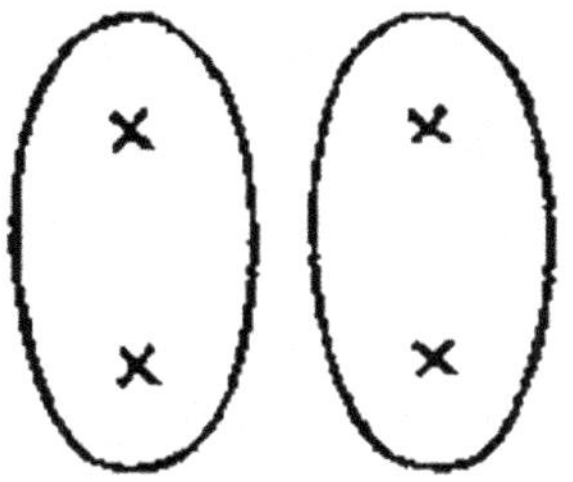

to see that 2 + 2 are 4."—Then I only need to look at the figure

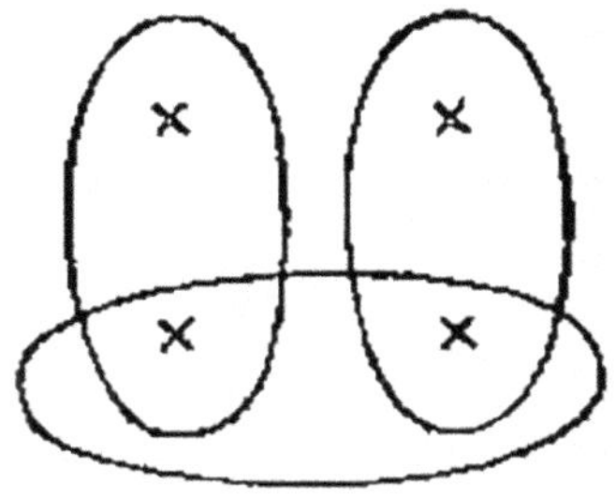

to see that 2 + 2 + 2 are 4.

(*RFM* I §38: 52)

The first picture appears to divide 4 into two pairs, but then the second picture presents 4 as *three* pairs. And Wittgenstein seems to suggest that while we would normally give normative force to the former, we could instead have stipulated that the latter was to be a proof that 2 + 2 + 2 = 4.

The response to this last example is that by proposing such a drawing of three pairs as a proof 'that 2 + 2 + 2 = 4' one would evince a rather different concept of addition. The three pairs overlap, whereas it is essential to *our* concept of addition that the amounts to be added are strictly distinct. The implication of Wittgenstein's example is that we have to imagine this 'proof' to be propounded *before* the advent of our calculus of arithmetic: when people have learnt to count, but are as yet experimenting with possible arithmetic operations. Then the diagrammatic 'proof that 2 + 2 + 2 = 4' would amount to introducing the operation of overlap-addition (already presented in Chapter 9). If we symbolize this operation by the obelus sign '†', we can define:

$$n \dagger m \quad = \quad n + m - 1$$

On this reading, it becomes readily plausible to say that the proof involves an element of stipulation, establishing a new concept of addition (overlap-addition).

The *H-P* proof, too, introduces new concepts. As already mentioned, Wittgenstein suggests that it fixes the meaning of the expression 'has the same number'. I'd rather emphasise that it introduces the new concepts of a Hand and a Pentacle. Merely being told that this is a Hand:

(a)

and this is a Pentacle:

(b)

—we cannot yet rule out that the following are not also to count as a Hand and a Pentacle respectively:

(d)

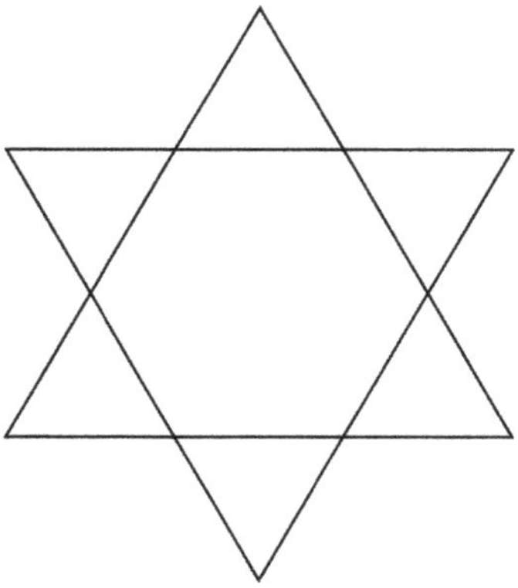
(e)

After all, we cannot be assumed to start with the understanding that a Hand has five strokes and a Pentacle has five corners, for then the proof would be superfluous. Hence, it is only the proof by correlation between (a) and (b) that tells us that the number of strokes is an essential feature of a Hand and that the number of angles is an essential feature of a Pentacle, and that they are equal.

Finally, how can the geometrical proof that four triangles can form a rectangle be said to involve an element of stipulation? Wittgenstein himself provides the incredulous question whether it is at all possible *not* to accept it:

> I said, "I accept such-and-such as proof of a proposition"—but is it possible for me *not* to accept the figure shewing the arrangement of the pieces as proof that these pieces can be arranged to have this periphery? (*RFM* I §48: 56)

His attempts to illustrate how one may fail to be convinced by that proof appear rather feeble. Of course it is conceivable that one 'tries all possible arrangements and always passes this one by, as if bewitched' (*RFM* I §44: 56); even after having been shown the proof. But why should that make me doubt the proof's conclusiveness? The proof makes no claim about people's ability to solve the puzzle, it only shows that a solution exists (and what it looks like). And that one cannot do it even after having seen it once is not surprising either: we often find ourselves unable to remember a trick we once knew.

Apart from envisaging a psychological difficulty ('a "blind spot" in our brain'), Wittgenstein also considers that the puzzle may remain unresolved due to 'the physical properties of the parts' (*RFM* I §44: 55–6). A piece may inadvertently have been turned upside down, presenting the mirror-image of the original shape, so that now it is impossible to complete the pattern (*RFM* I §49: 57). We could also imagine that owing to changes in temperature some of the pieces slightly alter their size or shape and then no longer fit the pattern.—Here again we encounter different concepts: the triangles in question can be taken as geometrical figures or as physical pieces of a certain material, shape, and size. To find the proof uncompelling for the reasons just suggested means: assuming an unsuitable concept of a triangle—as a changeable physical object, rather than a geometrical shape. Accepting the proof, on the other hand, means accepting a geometrical concept of a triangle, instead of the physical one that makes the solvability of the puzzle a mere contingency.

Another way in which Wittgenstein expresses the claim (iii) that a mathematical proof contains an element of stipulation is by saying that a *proof produces a new concept* (*RFM* 166b, 248b, 297f, 411b, 432f, 434cd, 435b) or *changes a concept* (*RFM* 166d).[2] Let us discuss the plausibility of this claim by considering a few real examples.

2. In most cases this may amount to the same idea, as one can say interchangeably 'this changes our concept of *X*' or 'this introduces a new concept of *X*', although talk of introducing new concepts is also applicable to cases where mathematicians come up with an entirely new construction.

(a) Proof That $a^0 = 1$

Alice Ambrose argued that Wittgenstein's view of proofs is convincing only in cases that should not really be regarded as proofs, although they are often presented under this title (1959). For example, when a mathematics textbook gives as a justification for $a^0 = 1$ the following:

$$a^m \div a^m \quad = \quad a^m/a^m \quad = \quad 1$$
$$a^m \div a^m \quad = \quad a^{m-m} \quad = \quad a^0$$

—this should indeed be regarded as giving meaning to the sign 'a^0', or introducing the new concept of exponent *0*. After all, according to the original definition of an exponent as the number of times a number is multiplied by itself, the expression 'a^0' would be meaningless. But then, it's not really a proof, but merely reasoning adduced to justify a convention. (The stipulation $a^0 = 1$ makes the formula '$a^n/a^m = a^{n-m}$' valid without restriction, even for $n = m$.) Whereas a deductive proof (such as a proof in Euclid) compels us to accept a theorem on pain of contradiction, mathematicians might well have decided not to accept that definition of 'a^0' (as, for example, Descartes refused to accept the conceptual innovation required for $x^2 = -1$ to have a solution) *without contradicting any of their previous assumptions* (Ambrose 1959, 441–3).

(b) Skolem's Inductive Proof of the Associative Law of Addition

It is probable that Wittgenstein's claim (iii)—that a proof involves an element of stipulation—was inspired by his reflections on inductive proofs, in particular, Skolem's proof of the associative law of addition (already considered in Chapter 4.2), which can be presented as follows (Waismann 1936, 91):

1.	$a + (b + 1) = (a + b) + 1.$	*Definition A(1)*
2.	$a + (b + c) = (a + b) + c$	*Assumption A(c)*
3.	$a + (b + [c + 1]) = a + ([b + c] + 1)$	*From 1*
4.	$= (a + [b + c]) + 1$	*From 2*
5.	$= ([a + b] + c) + 1$	*From 2*
6.	$= (a + b) + (c + 1)$	*From 1*
7.	$a + (b + [c + 1]) = (a + b) + (c + 1)$	*From 3–6: A(c+1)*
8.	If: $a + (b + c) = (a + b) + c$, then: $a + (b + (c + 1) = (a + b) + (c + 1)$.	*From 2–7: A(c) ⊃ A(c+1)*
9.	$a + (b + c) = (a + b) + c.$	*From 1 & 8, by complete induction: A(c)*

The base case *A(1)* is introduced by definition; then the induction step is proven: assuming *A(c)* we can derive *A(c + 1)*; hence: *A(c)* ⊃ *A(c + 1)*; and finally, base case and induction step are taken to prove the general algebraic law *A(c)*.

Initially, Wittgenstein had been inclined to deny that this was a proof of the associative law of addition, holding along the lines of the *Tractatus* that such a law could not be stated or proven as a proposition, but only show itself (*PR* 198, 203). More importantly, he pointed out that it was not properly derived as 'the end of the *chain of equations*' (*PR* 197); rather, it took a leap of reflection to get from base case and induction step to a generalisation. And in any case, the upshot was a basic law of a new calculus, algebra, and could as such only be stipulated, not proven (*PR* 193). But then, in the early 1930s, he changed his mind, accepting a broader notion of proof. And interestingly, the very feature of Skolem's proof that initially he had regarded as a reason to deny proof status—namely, that its conclusion is adopted as a new rule, making for a new, or at least modified, calculus—that very feature he now took to be an essential mark of a mathematical proof: claim (iii) above. Indeed, Skolem's proof justifies and establishes its general conclusion, but at the same time we take it as the beginning of the new calculus of algebra. Yet the fixation of the axioms of a new calculus can with propriety be said to be a matter of stipulation: of introducing a new conceptual system, which of course might have been defined differently.

An important aspect of this example is of course the way in which a representation of infinity is introduced here, when we take the step from arithmetic to algebra. Potential infinity, endlessness is a feature of our natural number system and hence of arithmetic. When we introduce complete induction and the symbolism of algebra we refer to that feature in a way that might be taken to make it an infinite totality. Wittgenstein was keen to emphasise that this amounts to the introduction of a new calculus, which, moreover, need not be understood to assume the existence of infinite totalities.

(c) Cantor's Diagonal Proof

Wittgenstein's discussion of Cantor's Diagonal Proof is based on his views on irrational numbers already developed in the early 1930s (in *PB* and *BT*), in particular his rejection of the idea of real numbers as infinite expansions. Rather, an irrational number, such as $\sqrt{2}$ or π is a rule or law for generating a series of rational numbers, decimal fractions:

> The idea behind $\sqrt{2}$ is this: we look for a rational number which, multiplied by itself, yields 2. There isn't one. But there are those which in this way come close to 2 and there are always some which approach

> 2 more closely still. There is a procedure permitting me to approach 2 indefinitely closely. This procedure is itself something. And I call it a real number.
>
> It finds expression in the fact that it yields places of a decimal fraction lying ever further to the right.
>
> (*PR* 227)

> A real number *yields* extensions [expansions], it is not an extension [expansion].
>
> (*PR* 228)

Irrational numbers are certain laws, not points to fill gaps on the number line. Wittgenstein rejects the idea of the continuum (*RFM* 286d). There are no gaps between the rational numbers (*BT* 754), to be filled by irregular infinite decimal fractions. The picture of points on a number line assimilates irrational to rational numbers, whereas Wittgenstein is keen to stress their heterogeneity: a rational number, which may be written as a (finite or recurring) decimal fraction, is obviously very different from a law for the construction of an open-ended series of such decimal fractions. This crucial difference is misleadingly hidden from view by the common fiction of an infinitely long (non-recurring) decimal expansion, when in truth all we have to play the role of an irrational number sign [*Zahlzeichen*] is the rule for developing the series [*Entwicklungsregel*] (MS 126, 137–8 [1942]).

This account of real numbers, developed in the early 1930s, is presupposed in Wittgenstein's 1938 discussion of Cantor's Diagonal Proof. That irrational numbers have no finite (or recurring) expansions simply means that the technique of calculating them has no limit, not that it results in an infinite expansion—there is no such thing (*RFM* 138c).

Why call irrational numbers 'numbers' at all, rather than 'series of numbers'? Because they can be the result of arithmetical operations (e.g. $\sqrt{x}$), we can further calculate with them, and they can be compared in size with rational numbers (determined to be smaller or greater than any given rational number) (*PR* 236).

By that criterion, however, lawless irrationals—e.g. expansions to be determined step by step by throwing a die—should not count as numbers (*PR* 229, *PG* 485).

More controversially, Wittgenstein also rejects the idea of defining an irrational number as the result of replacing certain digits in a regular irrational number, e.g. every '5' in the expansion of $\sqrt{2}$ is to be replaced by a '3': ${}^{5\rightarrow 3}\sqrt{2}$ (*PR* 224–5).

Why? After all, such irrationals derived from other irrationals by replacement of certain digits are generated by a recursive law and also 'effectively comparable with every rational number' (Marion 1998, 194–6). But consider the number π^* defined as follows: π^* *has the same* *expansion as π except that any row of 100 successive 7s is to be replaced by a 0 if it succeeds an even digit, otherwise it is to be replaced by an 8.* Since we don't know of any row of 100 successive 7s in the expansion of π, we cannot effectively decide whether π* is = π, < π or > π (cf. Davis & Hersh 1982, 372–3). Therefore, since real numbers should be effectively comparable in size not only with rational numbers, but also with other real numbers, we have a reason not to regard π* as a real number.

As a further requirement for a genuine irrational number Wittgenstein suggests the following:

> The sign π' (or: $^{7\rightarrow 3}\pi$) means nothing, if there isn't any talk of a 7 in the law of π, which we could then replace by a 3.
>
> (*PR* 228)

In other words, the replacement rule (in this case: 7→3) must apply directly to the initial rule (in this case: π) and not only to the expansion resulting from it (Da Silva 1993, 94).

However, this criterion appears to be based on Wittgenstein's occasional loss of nerve vis-à-vis the rule-following considerations, i.e. the idea, towards which he sometimes inclined in the early 1930s, that an expansion is not *determined* by a law (*PG* 481, 479; cf. Chapter 7 above).

At *PR* 235 he suggests more carefully that there may be determination, but 'I can't see in what way' the result is determined (e.g. the series of prime numbers). So in order to see if and where the replacement instruction kicks in one has to keep developing the underlying irrational number.—But why should that be a reason not to regard π' as a genuine irrational number?

Victor Rodych points out another criterion Wittgenstein puts forward for accepting something as a genuine irrational number (1999b, 287): a general rule of operation should be invariant across different base-notational systems, whereas the '7→3' replacement rule depends on the contingent fact that we use the decimal system (*PR* 231–2). 'In this way $^{7\rightarrow 3}\pi$ makes the decimal system into its subject matter'. Its application is not the expression of an arithmetical law, 'but only makes a superficial alteration to the language' (*PR* 232).

Cantor's famous Diagonal Proof is intended to show that the real numbers are uncountable (or non-denumerable). It is a reductio ad absurdum argument: one first assumes that all real numbers, say between 0 and 1, written as non-terminating decimal fractions, can be ordered, i.e. brought in one-to-one correlation with the natural numbers, and then shows how to construct a number that is not in this series: by stipulating that it differ from each number of the series in at least one digit: in its first digit it differs from the first number, in its second digit it differs from the second number, and so on. Hence, the real numbers cannot be one-to-one correlated with the natural numbers. There are, it would appear, *more* real numbers than natural numbers, or as Cantor (1891, 76–7) puts it: the infinite set of real numbers has a greater cardinality than the infinite set of natural numbers.

To begin with, Wittgenstein objects to the way Cantor's proof invites us to regard irrational numbers as infinite decimal fractions.

> The picture of Cantor's considerations is extremely misleading. For it shows us expansions—number signs that cannot actually be used as number signs. So it's not clear whether we have inserted a new expansion among the expansions (as it appears) or whether we've demonstrated a new law.
>
> (MS 178c, 1)[3]

Wittgenstein's view is that the diagonal method does *not* yield a new irrational *number*; just as he doesn't accept π' (the rule for replacing 7s in π by 3s) as a genuine number. Obviously, Cantor's 'diagonal number' is a similar kind of construction. The choice of certain digits in the expansions is dependent on the contingent choice of the decimal system for calculating the numbers listed. Moreover, the ordering of those numbers appears to be arbitrary,[4] which makes the resulting 'diagonal number' equally arbitrary, i.e. not determined by any mathematical rule.

Cantor's construction doesn't yield what Wittgenstein would be inclined to call a number, but merely an expansion, or more precisely: a recipe for the unlimited construction of an expansion.

> Cantor's diagonal procedure does not shew us an irrational number different from all in the system, but it gives sense to the mathematical proposition that the number so-and-so is different from all those of the system.
>
> (*RFM* 133f-134a)

That is to say, Cantor's procedure shows us the beginning of an *expansion* different from all the expansions of the real numbers in a given list (*RFM* 134d), not the *number* (i.e. a proper mathematical rule) that allows us to produce that expansion (cf. Han 2010). Such a number remains to be found:

> Thus it can be *set* as a question: Find a number whose expansion is diagonally different from those in this system.
>
> (*RFM* 134e)

3. *Das Bild der Cantorschen Überlgg. ist ungemein irreführend. Es zeigt uns nämlich Extensionen—Zahlzeichen, die doch nicht als Zahlzeichen zu benützen sind. So daß es nicht klar ist, ob wir den Extensionen eine neue Extension einfügen (wie es ja aussieht) oder ein neues Gesetz aufzeigen.*
4. In his 1891 article Cantor considers 'any [*irgend eine*] simply infinite series of elements of set M' (76).

To regard Cantor's procedure as the answer to that question would be like regarding the task 'Divide an angle into three' as carried out by laying 3 equal angles together (*RFM* 125d–126a). In this case, too, we have the *result* of the required answer (an angle subdivided into three equal angles), but not the answer (the procedure of subdividing a given angle). Likewise, Cantor shows us a trick to produce a certain kind of *expansion* without specifying anything that Wittgenstein would want to accept as a number (or mathematical law) to produce that expansion.

Again, the same suspicion that presenting the diagonal procedure as a new *number* is but a dubious trick is expressed by the following analogy:

> Would this not be as if any row of books were itself ordinarily called a book, and now we said: "It makes no sense to speak of 'the row of all books', since this row would itself be a book."
>
> (*RFM* 130c)

In other words, Cantor's proof depends on using the word 'number' in an extended sense. Note that here Wittgenstein does not object to that usage: assuming we 'ordinarily' apply the word 'book' also to a row of books suggests that regarding the diagonal decimal expansion as a number would also strike us as a natural use of the word 'number'. However, we should realise that it involves a conceptual extension.

Rodych observes that Wittgenstein's rejection of lawless and even 'digit replacement' irrationals as proper numbers is an instance of essentialism, and hence quite contrary to Wittgenstein's professed philosophical approach (1999b, 298–9). But the categorical rejection of such numbers seems to occur only in Wittgenstein's middle-period writings (*PR*, *BT*), while the 1938 remarks published as Part II of *RFM* are less dogmatic. Although in some remarks Cantor's construction is said not to be (what Wittgenstein would call) a number, in others Wittgenstein is more concerned to *point out* different uses of the word 'number' (cf. *LFM* 15), and in some cases to question their usefulness, rather than dogmatically banning them as mistakes. After his criticism of the diagonal number he considers the reply that, after all, it fulfils the criterion of being comparable in size to other numbers and we can calculate with it (*RFM* 126c). He responds:

> Is the question not really: What can this number be *used* for? True, that sounds queer.—But what it means is: what are its mathematical surroundings?
>
> (*RFM* 126d)

Cantor's proof involves a new concept of a number (based on a rule for digit replacement in the decimal system) whose mathematical usefulness remains to be seen. Thus, it appears to confirm Wittgenstein's claim that mathematical proofs involve conceptual change.

Another line of criticism appears to be that it is not clear what Cantor shows to be impossible (*RFM* 129c, 130bc, 132c, MS 126, 10–12). Of course we could list all irrational numbers that spring to mind, but there would be no reason in the first place ever to think that we've listed them all.—However, even though, as noted above, Cantor appears to present the initial series of real numbers as arbitrarily chosen, modern accounts would offer something systematic: some function from natural numbers to real numbers. In any case, it doesn't seem to be a strong objection to an impossibility proof that what is shown to be impossible is initially described in a vague or ultimately—as the proof shows—incoherent way.

More importantly, Wittgenstein also objects to Cantor's taking the result of his proof to say something about the set of *all* real numbers:

> If you call Cantor's procedure one of producing a new real number you will now no longer be inclined to speak of a system of all real numbers.
>
> (MS 121, 71)[5]

Granted that Cantor succeeds in showing that no proposed ordering of real numbers can be regarded as complete because we can always construct a number not covered by the ordering, why should we then feel entitled to speak of a complete set of real numbers? Isn't the upshot of the proof rather the opposite: that we have no concept of a totality of real numbers? After all, it's not as if the initial ordering plus the diagonal number would be the totality of real numbers (cf. Goodstein 1972, 275). Rather, whatever diagonal numbers we add to a given ordering, it would still not be a complete set. Hence, for Cantor to draw a conclusion about the cardinality (*Mächtigkeit*) of the set of all real numbers strikes Wittgenstein as rather misleading: as it makes the 'difference in kind' between rational and real numbers appear to be merely a 'difference of extension' (*RFM* 132d), i.e. a difference in the size or cardinality of the set.—Another way of expressing Wittgenstein's concern is this. What gives sense to our talk of 'all natural numbers' is induction: the possibility of showing that what holds for one member of a sequence also holds for the next; then, if it holds for the first we can say that it holds for *all* of them. But induction requires a complete ordering such that any particular number will be reached if you go on for long enough. As Cantor's Diagonal Proof shows the impossibility of such a complete ordering it gives us a reason *not* to speak of 'all real numbers', in the sense in which

5. *Wenn Du nun das Cantorsche Vorgehen eines nennst, eine neue reelle Zahl zu erzeugen, so wirst Du nun nicht mehr geneigt sein, von einem System aller reellen Zahlen zu reden.*

we *can* speak of all natural numbers or all real numbers of a certain kind, e.g. all square roots.[6]

Furthermore on Cantor's conclusion, Wittgenstein comments as follows:

> It means nothing to say: "*Therefore* the X numbers are not denumerable." One might say something like this: I call number-concept X non-denumerable if it has been stipulated that, whatever numbers falling under this concept you arrange in a series, the diagonal number of this series is also to fall under that concept.
>
> (*RFM* 128b)

That is to say, the result of the proof is the determination of a new concept of non-denumerability (or uncountability). Our ordinary concept of countability (of what can be counted) is trivially inapplicable to infinity—for that is what 'infinity' means: that counting will not yield a result (cf. Shanker 1987, 196). Obviously, such ordinary countability is not at issue here, for according to Cantor the infinity of natural numbers *is* denumerable. Rather, the idea is that we have an ordering of numbers of a certain kind such that we will reach any given number of that kind if we just extend the series far enough. That is obviously the case with natural numbers and Cantor presented such an ordering for rational numbers (cf. *RFM* 140c). For real numbers, however, we have no such ordering and Cantor's Diagonal Proof shows, as Wittgenstein acknowledges, that it is futile to continue looking for one (*RFM* 129c). For any assumed system of ordering we could construct a number not covered by it, thus showing that it is not a proper system of ordering after all. And that is what according to Cantor's proof the new concept of non-denumerability means.

What Wittgenstein sets his face against is the view that Cantor's Diagonal Proof shows that there are *more* real numbers than natural numbers. On this picture, a non-denumerable set is simply bigger than a denumerable one: so big that we cannot even begin to count it (because it cannot be comprehensively ordered). As if we already had a concept of size applicable to both finite and infinite sets, allowing us to distinguish between different sizes of infinity (*RFM* 132d; *PG* 464). The idea would be that by the relative size of two sets we simply mean something to be decided by the criterion of one-to-one correlation: if they can be one-to-one correlated they are of equal size, if when all members of one set have been correlated there are still uncorrelated members of the other set, then the other set is bigger. Our concept of size of set, based on the criterion of

6. I owe this point to David Dolby.

one-to-one correlation, would appear to be applicable both to finite and infinite sets, as Cantor demonstrated.

However, it is not true that the criterion of possible one-to-one correlation exhaustively characterises our ordinary concept of the relative sizes of sets. Another criterion, just as intuitive, is the part–whole principle according to which a proper subset must be smaller than the set that contains it and more. But when it comes to infinite sets, those two ordinary criteria come apart. Although natural numbers and even numbers can be one-to-one correlated, it remains overwhelmingly plausible to say that there are *more* natural numbers than even numbers: for the even numbers are a proper subset of the natural numbers. Cantor accepted the one-to-one correlation criterion as the criterion of size of infinite sets, too. But the paradoxicality of his claim that a set can have the same size as its proper subset (often illustrated by the story of Hilbert's Hotel) shows that this is no longer our ordinary concept of size of sets, as applied to finite sets. And Paolo Mancosu (2009) has shown that Cantor's way of extending the concept of size to infinities was not the only conceptual option. Not only could one refrain from quantifying infinite sets because, holding on to our ordinary concept, that leads to contradiction, as Galileo and Leibniz maintained.[7] There have also been some recent mathematical developments (Katz, Benci, Di Nasso, Forti) that generalize not the one-to-one correlation principle, but the part–whole principle to infinite sets in a coherent fashion (Mancosu 2009, 639–42).[8]

To conclude, Cantor's Diagonal Proof appears to be a plausible case in support of Wittgenstein's claim that mathematical proofs involve conceptual change (cf. MS 163, 55r). First, there is reason to say that regarding a diagonal ('digit replacement') number as an irrational number amounts to an extension of the concept of an irrational number. Secondly, we don't seem compelled to speak of a set of *all* real numbers. Finally, Cantor's introduction of different cardinalities of infinite sets seems to be a conceptual change to which, as has been argued, there would have been reasonable alternatives.

(d) Euclid's Construction of a Regular Pentagon

Euclid's proofs that a certain kind of geometrical construction is possible may be thought to provide another suitable example to support

7. Galileo 1632, 32–33; Leibniz 1875–1890, 315 (quoted in Mancosu 2009, 618).
8. Contrary to Gödel's claim that Cantor's development was inevitable (Gödel 1947, 259), which presupposes that we regard infinite developments as complete sets (Mancosu 2009, 637–9). Not only are we not compelled to do so, but it is arguably an inconsistent idea, as illustrated by Wittgenstein's scenario of 'a man whose life goes back for an infinite time and who says to us: "I'm just writing down the last digit of π, and it's a 2"' (*PR* 166).

Wittgenstein's account of proofs as involving conceptual change. In Book IV, Proposition 11, of the *Elements*, Euclid demonstrates how it is possible to construct a regular pentagon inscribed in a circle, using only compass and ruler. Following Wittgenstein, one might suggest that this amounts to defining the very concept of a regular pentagon. Felix Mühlhölzer argues for this view on the grounds that in Euclid's conceptual framework there is no measurement, hence no criterion for equal length apart from a construction. Therefore, without the construction proof one cannot even make sense of the definition of a regular pentagon as a figure with five sides of equal length and five equal angles (Mühlhölzer 2001, 229–30).

However, the equal lengths of two sides can be established even without a marked ruler or measure, by means of a compass. After all, the principle of a compass is that from a given point one traces *the same distance* in all directions. Thus, in Euclid's very first Proposition (I.1), for example, an equilateral triangle is constructed by using a compass to trace the length of the base line AB first from A and then from B in various directions, and where the two intersect we have a point C, such that both AC and BC are of the same length as AB.

Hence, even without a construction proof we can easily explain what we mean by a regular pentagon (indeed, by a regular *n-gon* for any number *n*) by giving the following *practical* instruction on how to draw it by trial and error:

> Draw a circle, choose a point *A* on that circle, fix a length between the arms of a compass smaller than the circle's diameter, then, going clockwise, find a second point *B* on the circle at the fixed distance from *A*, then a third one *C* at the same distance from *B*, then a fourth one *D* at the same distance from *C*, and similarly a fifth one *E*, and a sixth one *F*. If *A* = *F*, the lines between the points will form a regular pentagon. If, however, on your clockwise tour along the circle *F* hasn't got round to *A* yet, increase the length and try again. If, on the other hand, by reaching *F* you have gone beyond *A*, the length must be decreased. Keep changing the length until *A* and *F* coincide.

Obviously, an analogous instruction can be given for drawing and identifying a heptagon, even though a Euclidian construction of a heptagon is impossible (as shown by Gauß in the 19th century). Therefore, I think it is implausible to say that without the constructive proof we cannot even understand the concept of an equilateral equiangular figure with 5 (or indeed 7) sides.

More plausible, then, is the claim that the proof introduces the concept of a *construction of a pentagon with compass and ruler*. But is that true? It might be objected that such a claim confuses the concept with something falling under the concept. That I don't know who the tallest man alive is (not even whether there is *one* man who is the tallest) doesn't mean that I haven't got the concept. Similarly, it might be suggested that since we know what a pentagon is and we understand the idea of a systematic procedure of drawing a figure using only compass and ruler, we *ipso facto* have the concept of a *construction of a regular pentagon with compass and ruler*—even while we don't know how to produce an instance of it.

However, although it is true that possession of a concept does not require us to produce an instance of something falling under the concept, it does require that, logically, something *could* fall under it. For an expression to denote a concept it must be consistent. Thus there is no concept of a round square, as the expression 'round square' is self-contradictory. In mathematics, an investigation of whether a newly formed predicate has an application is quite unlike an empirical investigation, for example, as to whether there are any flying fish. There is obviously nothing inconsistent in the idea of an animal that can both swim under water and fly, but as long as we haven't found a construction of a regular pentagon, we do not know whether it is *logically* possible or logically impossible (as the construction of a heptagon). But it is plausible to say that as long as we don't even know whether a newly formed predicate '*F*' is consistent—whether we can make sense of '*F*'—we cannot really claim to have mastered, or possess, the concept of an *F*. Therefore, in the case of a Euclidian construction proof, we can indeed agree with Wittgenstein's view that it provides us with a new concept.

(e) Euclid's Proof That There Is No Greatest Prime Number

In Book IX §20 of the *Elements*, Euclid offers a reductio ad absurdum proof that there is no greatest prime number, which can be paraphrased as follows:

> Suppose P_n is the greatest prime number, then multiply all the prime numbers up to P_n and add 1:
>
> $(P_1 \times P_2 \times P_3 \times \ldots \times P_n) + 1$
>
> The resulting number is either prime, in which case P_n isn't the greatest prime after all; or, if it isn't, it must be divisible by a prime number, but it isn't divisible by any prime number up to P_n (for there will always be rest *1*), so it would have to be divisible by a greater prime number, in which case, again, there is a greater prime number than P_n. So there cannot be any greatest prime number P_n.

Does this proof introduce a new concept or change an existing one?

The most obvious candidate of a *new* concept involved in this proof is that of the greatest prime number, yet that concept is not introduced, but rejected: ruled out as inconsistent. Does it change our concept of a prime number to learn that there is no greatest prime number? Does it not merely enrich our knowledge of a given concept?

Moreover, we should not want to say that Euclid *stipulates* that the expression 'greatest prime number' have no application; rather, he shows that that is a conclusion forced upon us as it can be deduced from our concept of a prime number (as divisible only by 1 and itself). Similarly, it appears implausible to say that with this proof Euclid *changes* the concept of a prime number. Surely, the concept remains the same; we are only shown one of its implications. Indeed, had Euclid simply *stipulated* a new concept of a prime number, rather than convinced us of an implication of our original concept, we should not call that a proof. That appears to be Wittgenstein's thought in the following passage:

> As a *proof*, I might say, it has to convince me of something. In consequence of it I will do or not do something. And in consequence of a new concept I don't do or not do anything.
>
> (*RFM* 298b)

To *stipulate* a new concept is not to convince anybody of anything. For a concept is not a fact (*RFM* 298e). However, Wittgenstein wrote those remarks only tentatively, considering immediately afterwards how they might be rejected, suggesting that a concept could, after all, be said to convince me of something, namely: that I want to use it (*RFM* 298e). Yet elsewhere he envisaged conviction not to be occasioned by a stipulation, but, conversely, a conviction leading to a new stipulation:

> Let us remember that in mathematics we are convinced of *grammatical* propositions; so the expression, the result, of our being convinced is that we *accept a rule*.
>
> (*RFM* 162d)

> For a mathematical proposition is the determination of a concept following upon a discovery.
>
> (*RFM* 248b)

This is to be a two-step process: *first* we are convinced of something (or make a discovery), *then* we change our concepts accordingly.[9] I suppose what Wittgenstein had in mind here is something like the geometrical

9. The two steps reflect the two phenomenological aspects of mathematics Wittgenstein tries to reconcile: 'I should like to be able to describe how it comes about that mathematics appears to us now as the natural history of the domain of numbers, now again as a collection of rules' (*RFM* 230e). That is the 'twofold character of the mathematical proposition—as law and as rule' (*RFM* 235e), i.e.: factual and normative. It serves as a rule when applied outside mathematics, while in the realm of mathematics it appears more like a fact-stating 'natural law' (MS 127, 236). This is, again, the combination of the grammar view and the calculus view.

puzzle proof (of *RFM* I §42): where something is discovered only in a particular case, but then generalised in such a way that the generalisation determines a concept (in that case of a triangle as a geometrical shape). Elsewhere Wittgenstein writes:

> One would like to say: the proof . . . changes our concepts. It makes new connexions, and it creates the concept of these connexions.
> (*RFM* 166d)

First we find a new connexion, *then* we fasten it in a new concept. This means that derivation and conceptual stipulation must be logically distinct. It should be conceivable, on this account, that one occurs without the other. That the latter occur without the former is easily conceivable: a mathematical concept can be changed or introduced even without a preceding derivation. There are indeed historical examples of this kind: rules of thumb that were conventionally adopted without any proof. For instance, the Babylonian rule for calculating the area of a circle, based on approximate measurements (Meschkowski 1984, 41). However, if mathematical proof is really such a two-step affair, it should also be possible to have only the first part: a derivation whose results are then not adopted as theorems.

Earlier, Wittgenstein's view that mathematical propositions are rules produced according to other rules was compared to a legal system in which the passing of new laws is regulated by procedural laws. In that case it is indeed possible that the result of some legal procedures is then not adopted. Thus, in 1895 the Viennese City Council elected Karl Lueger to the position of mayor; yet the Emperor Franz Josef I. refused to approve of the election and so, on that occasion, Lueger did not become mayor of Vienna. Here we have two distinct procedures: the election and the imperial confirmation. Although the latter is only to give a seal of approval to the former, that may not happen, as seen in Lueger's case. Instead of a lacking imperial endorsement of the result of some legal procedures, we could also imagine a popular refusal to accept that result. Thus, a correctly passed law may be generally ignored, and not enforced by the police, and so never become operative. Is something like that conceivable in the mathematical case?

It doesn't seem to be a correct description in the current case. Once Euclid had produced his reasoning to show that there cannot be a greatest prime number, no further endorsement was necessary for it to become a mathematical proof. No committee of leading mathematicians was called upon to decide whether to put that proof into an archive of valid mathematics. As argued above in Chapter 8, talk of such an archive is only a metaphor for mathematical results that are readily remembered and so need not be calculated again and again (such as elementary multiplications memorised at school). Yet what is *not* generally known is not for that matter barred from being mathematics.

When Euclid had constructed his proof he did not have to wait for his colleagues' approval for it really to *become* a proof. Again, when I encounter the proof for the first time I can convince myself that it *is* a proof by checking the conclusiveness of the reasoning; I do not have to carry out any sociological research to see if the proof is widely accepted. Popular or expert acclaim is not a criterion of correctness in mathematics.

At one point at least Wittgenstein seems to anticipate that objection:

> But what if someone now says: "I am not aware of these *two* processes, I am only aware of the empirical, not of a formation and transformation of concepts which is independent of it; everything seems to me to be in the service of the empirical"?
>
> (*RFM* 237f)

By 'the empirical' he appears to mean what mathematicians *find* to be the case, rather than stipulating it; as if he hadn't noticed that in the conceptual, a priori, realm, too, one can make discoveries. At any rate, that objection is left unanswered.

Nonetheless, repeatedly, in Wittgenstein's discussions of mathematical proof one encounters passages that seem to show an inclination towards the 'decisionist' account of rule-following, already considered above (Chapter 7), that is, the idea that each new application of a rule requires a decision. He writes, for example:

> [Accepting a proof:] I decide to see things like *this*. And so, to act in such-and-such a way.
>
> (*RFM* 309g)

This may be a careless overstatement, especially considering other more tentative remarks, which only record an *inclination* to put things in such a provocative manner:

> I want to say: it *looks* as if a ground for the decision were already there; and it has yet to be invented.
>
> (*RFM* 267e)

Or even *questioning* whether such a claim would be correct (note the double question marks at the end):[10]

> Can I say: "A proof induces us to make a certain decision, namely that of accepting a particular concept-formation"??
>
> (*RFM* 238f)

10. Missing in the English edition.

Other remarks make it, again, quite clear that the word 'decision' is in fact rather misleading. What inclines Wittgenstein to speak of a 'decision' is just the observation that reasons come to an end:

> Doesn't its being a spontaneous decision merely mean: that's how I act; ask for no reason!
>
> You say you must; but cannot say what compels you.
>
> (*RFM* 326cd)

Yet what makes it rather misleading to speak of a 'decision' here is that, as Wittgenstein acknowledges, I have no choice.

> When I say "I decide spontaneously", naturally that does not mean: I consider which number would really be the best one here and then plump for . . .
>
> (*RFM* 326f)

> "As I see the rule, *this* is what it requires". It does not depend on whether I am disposed this way or that.
>
> (*RFM* 332e)

If one wanted, in such a case, to speak of a 'decision', really, one would have to add that one *must* decide this way (*RFM* 326b). For in spite of his recurring inclination to use the word 'decision', Wittgenstein does not deny that I am *compelled* to take the steps I take:

> "The rules compel me to . . ."—well, one can indeed say that, for after all it is not a matter of my own will what seems to me to agree with the rule.
>
> (*RFM* 394b; my translation)

In fact, compulsion is an essential feature of proofs and rule-following in general:

> If a rule does not compel you, then you aren't *following* a rule.
>
> (*RFM* 413d)

And yet, in another sense of the word, one may remonstrate that we are not *compelled* to accept a proof: it is after all conceivable that—like the deviant pupil in the rule-following discussion (*PI* §185)—we do not accept a given proof, thus manifesting a different (probably bizarre) understanding of some of the concepts involved. In as much as it is conceivable that we react in some such anomalous way, we are not compelled to react the way we do. Yet of someone who reacted in such a deviant way we should say that he had a different concept. Thus, what makes us feel compelled is our adherence to a given understanding of the concepts involved (*RFM* 328i–329a). True, nothing actually *forces* us to have such concepts, but we do: we are 'happy to accept this chain (this figure) as a *proof*' (*RFM* I §33: 50).[11] We can say then that we *must* accept the proof, because we don't want to deviate from the concept in question, as we understand it:

> For the word "must" surely expresses our inability to depart from *this* concept. (Or ought I to say "refusal"?)
>
> (*RFM* 238d)

In short, contrary to what Wittgenstein may have been inclined to say in some reckless or hyperbolical remarks, the rule-following considerations cannot plausibly be invoked to support the view that every proof introduces a new concept. On the contrary, the very idea of following a rule (or applying a concept) presupposes that the rule (or concept) remains the same from one application to the next.—

Mühlhölzer suggests that Euclid's proof could be said to 'create the concept of a transition from "$p_1, \ldots, p_n$" to "$p_1 \times \ldots \times p_n + 1$"' (2010, 242). It shows us how on the basis of a given series of prime numbers we can always construct a new and greater prime number:

> The proof teaches us a technique of finding a prime number between p and $p! + 1$. And we become convinced that this technique must always lead to a prime number $> p$.
>
> (*RFM* 307g–308a)

But is that new technique appropriately described as a 'new concept'?

As Wittgenstein remarked, 'concept' is a rather vague term (*RFM* 412g). To say that *one's* concept has been changed might just mean that one's understanding of a given concept has changed, which is undoubtedly true in this case. When I acquire a better understanding of the functioning of a computer, I can say that my concept of a computer has changed, which does not mean that *the* concept of a computer has changed.[12] Similarly, learning how I can always construct a greater prime number and thus realising that there is no greatest prime can be said to enrich 'my concept', i.e. my understanding of a prime number.

11. *D.h., ich lasse mir diese Kette (diese Figur) als Beweis gefallen.*
12. This is probably more obvious in Wittgenstein's German. The German word for 'concept': '*Begriff*', is derived from the common verb '*begreifen*': to grasp, to understand. And '*einen klaren (oder unklaren) Begriff von etwas haben*' (to have a clear [or unclear] concept of something) means: to have a clear (or unclear) understanding of something.

Or again, sometimes Wittgenstein stresses that a proof changes my 'way of seeing' (*RFM* 239ef), or gives me a 'clearer idea' (*LFM* 88), which are also a fairly uncontroversial claims.

And at one point he simply stipulates that what he means by 'concept' can be a method:

> He has read off from the process, not a proposition of natural science but, instead of that, the determination of a concept.
>
> Let concept here mean method.
>
> (*RFM* 310bc)

But a new method can perhaps also be described as a new way of handling some already existing concepts:

> Mathematics teaches us to operate with concepts in a new way. And hence it can be said to change the way we work with concepts.
>
> (*RFM* 413a)

That sounds unobjectionable: Euclid's proof gives us a method of producing ever greater prime numbers; not exactly a new concept, but a new technique involving our familiar concept of a prime number.

Wittgenstein's concept-change claim (iii) appears more plausible in the case of some other impossibility proofs: based on connections between different areas of mathematics (Goodstein 1972, 277–8). Thus, the impossibility of trisecting an angle with ruler and compass could be shown—but not within the domain of elementary geometry where the problem originated. The impossibility proof, produced by Pierre Wantzel in 1837, was based on transforming the geometrical problem into an algebraic one (of constructing a segment whose length is the root of a cubic polynomial). And it seems indeed plausible to say, with Wittgenstein, that such a correlation of geometrical concepts with algebraic ones amounts to giving the former a new, extended meaning. Again, Andrew Wiles's proof of Fermat's last theorem (that no three positive integers a, b, and c satisfy the equation $a^n + b^n = c^n$ for any integer value of n greater than 2) is an impossibility proof based on forging connections between—and thus enriching—distinct mathematical concepts (in particular, the concepts of elliptic equations and modular forms) (Singh 1997, 204–5).

(f) Proof (Calculation) in Elementary Arithmetic

For example:

25 × 25
50
<u>*125*</u>
625

This is the most humdrum type of mathematical proof: a simple arithmetical calculation. It is often referred to by Wittgenstein for illustrating another line of reasoning to the effect that proofs introduce new concepts. In various places the new technique introduced by a proof is described as that of forging and applying a *connection* between concepts, which Wittgenstein is inclined to regard as a new concept (*RFM* 297f). Although he evinces some uncertainty on that point:

> An equation links two concepts; so that I can now pass from one to the other.
>
> (*RFM* 296c)

> An equation constructs a conceptual path. But is a conceptual path a concept? And if not, is there a sharp distinction between them?
>
> (*RFM* 296d)

In the simplest case the proof may be a calculation resulting in an equation, e.g.: $25 \times 25 = 625$. Once calculated, the equation can serve as a substitution rule:

> Imagine that you have taught someone a technique of multiplying. . . .
>
> Now he says that the technique of multiplying establishes connexions between the concepts. . . .
>
> Now will he also be inclined to say that the process of multiplying is a concept?
>
> (*RFM* 296efg)

> Is a new conceptual connexion a new concept? And does mathematics create conceptual connexions.
>
> (*RFM* 412d)[13]

The idea appears to be that the fusion of two concepts, each defined by distinct criteria, results in a new concept. For instance, if initially by 'acid' we mean what *turns litmus paper red*, and then proceed to add to the definition of an acid that it *liberates hydrogen ions in water*, we have created a new concept of an acid. Similarly, 25×25 and 625 are initially distinct concepts, having different criteria of application (counting 25 groups of 25 members each, or counting 625). Then, the result of the multiplication entitles us to use them interchangeably. Thus, by connecting the two different criteria we have created a new complex concept: that of being both 25 × 25 and 625.

However, there is a crucial difference between the two cases. When we define a new concept of acid as what fulfils both criteria (*turns litmus*

13. The remark appears in the wrong place in the English edition. It should have come after *RFM* 412f: 'Now ought I to say . . .' (MS 124, 147).

paper red and *liberates hydrogen ions in water*) we have to check two things before we can declare something to be an acid. The new concept of an acid results from a conjunction of criteria. By contrast, the point of '*25 × 25 = 625*' is that henceforth the two criteria can be used interchangeably: having found something to be 25 sets of 25 items we *no longer* need to check that there are 625 items. In other words, the new 'concept' is not produced by a conjunction of criteria. On the contrary, it seems to be a merging of criteria. The proof, one could say, convinces us of their identity: by checking that something is 25 × 25 we have *ipso facto* checked that it is 625.

Wittgenstein invites us to see an equation as a two-sided concept. If one side applies, the other side applies as well. Consider a legal analogy. A law lays down that the *President* of the country is automatically the *First Lord of the Admiralty*. That law can be said to create a new concept: that of the joint position of President and First Lord of the Admiralty. However, these are proclamatory concepts: being proclaimed President (in the legally correct way), one becomes President. And so it is easy to see how the performative act of such a proclamation can be extended to make one First Lord of the Admiralty as well. Numbers, by contrast, appear to be descriptive, rather than proclamatory concepts. The number of apples I have cannot be determined by a proclamation; it is an empirical fact. So it cannot be a matter of proclamation or fiat either whether a given number is the same as the product of two other given numbers. And of course such a proclamation is also ruled out by the idea of proof.

The natural response to Wittgenstein's suggestion is to say that the link between *25 × 25* and *625* does not need to be created since it is already implicit in those concepts, as an internal relation between them (cf. Glock & Büttner 2018, 190–2). It is analytic that *25 × 25 = 625*. We may well say that *25 × 25* and *625* are 'two sides of the same concept' (*RFM* 297c), but not because of a decision to create such a concept.

Wittgenstein emphasises that the two sides of an arithmetic equation have different criteria of application. They do not say the same thing.

> How can you say that ". . . 625 . . . " and ". . . 25 × 25 . . . " say the same thing?—Only through our arithmetic do they *become one*. (*RFM* 358b)

That is indeed plausible. In our system of arithmetic we can show, step by step, that the two expressions are equivalent. But as Wittgenstein says in the following manuscript remark,[14] it is 'the *system* of arithmetic' that makes them equivalent, not the *particular proof* or calculation. As Wittgenstein suggested in 1933, a proof gives a new sense to a mathematical

14. *Erst als Glieder des Systems der Arithmetik werden sie eins.* [MS 161, 42r]

proposition in as much as it 'incorporates [it] into a new calculus' (*BT* 631), but by calculating that *25 × 25* yields *625* I do not develop a new calculus; I merely employ an existing one.

To summarise our findings so far, Wittgenstein's claim that *every* mathematical proof involves conceptual change does not appear to be tenable. In fact, Wittgenstein himself observed at one point that his concept-change claim should not be generalised:

> Here one should not want to be dogmatic: Of some new proof one will be inclined to say that it changes our concept, of some—so to speak, trivial one—not.
>
> (MS 163, 55rv; cf. 59rv)[15]

And yet he continues being inclined to regard concept change as an essential feature of proofs, without any qualification. Even of the most trivial proofs—arithmetical calculations or calculating some of the extension of an irrational number—he elsewhere suggests that they involve concept change (*RFM* 267d, 432f). So in spite of his awareness of exceptions to his claim,[16] he apparently continued to hold a somewhat exaggerated view of its applicability.

Proof and Experiment

If an open-minded consideration of a few examples of mathematical proofs does not in fact suggest that all proofs involve conceptual change, why should Wittgenstein have thought so in the first place? Is it perhaps his concern with the distinction between proofs and experiments that misled him? Consider the following remark:

> Imagine you have a row of marbles, and you number them with Arabic numerals, which run from 1 to 100; then you make a big gap after every 10, and in each 10 a rather smaller gap in the middle with 5 on either side: this makes the 10 stand out clearly as 10; now you take the sets of 10 and put them one below another, and in the middle of the column you make a bigger gap, so that you have five rows above and five below; and now you number the rows from 1 to 10.—We have, so to speak, done drill with the marbles. I can say that we have unfolded properties of the hundred marbles.—But now imagine that this whole process, this experiment with the hundred marbles, were

15. *Hier darf man nicht dogmatisch sein wollen: Von manchem neuen Beweis wird man zu sagen geneigt sein, er ändere unseren Begriff, von manchem—sozusagen trivialen—nicht.*
16. Cf. the concessionary wording of *RFM* 172b.

> filmed. What I now see on the screen is surely not an experiment, for the picture of an experiment is not itself an experiment.—But I see the 'mathematically essential' thing about the process in the projection too! For here there appear first a hundred spots, and then they are arranged in tens, and so on and so on.
>
> Thus I might say: the proof does not serve as an experiment; but it does serve as the picture of an experiment.
>
> (*RFM* I §36: 51)

This and other remarks (notably the simplified examples discussed at the beginning of this chapter) illustrate Wittgenstein's perspective on proofs. He considers how *empirical* support for an applied mathematical proposition—in this case an application of the equation '*10 × 10 = 100*'—can gain the status of a proof. How do we have to look at an experiment for it to become a proof? The answer is that we have to regard it as a mere *picture* of an experiment, taking its result as a criterion for the correct application of the rules (*RFM* 160f–161a, 319de–320a). And the step from seeing an experiment to taking it thus as a norm of representation can plausibly be regarded as a transition from the observation of a particular case to the formation of general concept—the two-step procedure mentioned earlier. This approach to proofs is still informed by the empiricist conventionalist view discussed in the previous chapter: the view of mathematics as a system of empirical observations 'hardened into rules'. As argued, that approach neglects the prominent role of proofs in mathematics. Wittgenstein, on the whole, does not. Never having abandoned his earlier calculus view of mathematics, he spends a lot of time discussing the nature of proofs. Yet it appears that his account remains unfortunately biased by those empiricist conventionalist ideas. Thus he remains inclined to regard proofs as originating from initial experiments (*RFM* 160c), which doesn't seem to be a suitable model for many, if not most mathematical proofs. Rather, a proof (or calculation) in arithmetic is not based on empirical instances (such as rows as marbles), but entirely on our concepts of numbers and operations that are already understood to have their full abstractness and generality. Such proofs occur at a level of systematic abstraction when the rudiments of mathematical thinking that fascinated Wittgenstein—where what is shown with concrete objects is first given a general, mathematical significance—have already been left behind.

However, even if the genetic account—that experiments are transformed into proofs—does not appear tenable, the underlying idea that the same occurrence could be regarded in two different ways (as an experiment or as a proof) remains very plausible. Perhaps the quasi-historical presentation (in remarks such as *RFM* 160c) should merely be taken as a picturesque way of juxtaposing the two opposing attitudes, as a kind of simplified developing language game (cf. *PI* §41).

10.2 What Is the Relation Between a Mathematical Proposition and Its Proof?

As explained earlier, Wittgenstein's mature philosophy of mathematics tries to combine two fundamental ideas which I called the *calculus view* and the *grammar view*. That is to say, mathematical propositions have two key features: definitions apart, they have to be *legitimised by proof* in order to be given the status of a *norm of representation*, akin to that of a grammatical proposition.

If truth in mathematics means that something has roughly the status of a grammatical rule, then it would appear that there cannot be any unknown mathematical truths. For the idea of a grammatical rule that nobody knows to be valid is nonsense: 'A law I'm unaware of isn't a law' (*PR* 176). For something to have a normative function it must be known to have a normative function; it cannot fulfil this function secretly.

However, this needs to be qualified. A system of rules may be so complicated that it's not always immediately clear whether a given move is correct or not. Legal systems are often so complicated that it takes an expert lawyer to work out how a given legal question is to be decided: what in this case would be in accordance with the law. However, for the legal system to function as such it must generally be possible, at least in principle, to work out what is and what is not in accordance with the law. Similarly, where a system of grammatical rules (e.g. for the use of addition, multiplication, etc.) allows for the generation of more specific grammatical rules (e.g. equations), it is only to be expected that their correctness may not always be self-evident. The crucial point is that it must always be possible to work out, to ascertain according to accepted rules, whether a given proposition is a grammatical rule or not.

If a given proposition is to be regarded as part of a system of norms, it must be such that we know how to check its status. Hence, Wittgenstein says of mathematical propositions: 'Every proposition is the instruction [*Anweisung*] for a verification' (*PR* 174).[17] In other words, a mathematical proposition (such as an arithmetical equation) comes with certain methods of proof, which belong to the proposition's mathematical sense.

An equation, such as '*25 × 25 = 625*', is not an isolated proposition, but part of a calculus (*PG* 376). Hence, the equation is not just a substitution rule (*WVC* 158; *PG* 347), such as the *isolated* rule '*12 × 12 = 144*' considered in Chapter 9; it also says that I get *625* if I apply the rules of multiplication to *25 × 25*. That is to say, the sign '=' has essentially two aspects. It means both 'can be replaced by' and 'yields according to the rules of arithmetic' (*PG* 377–8). In this way an equation refers to the calculation from which it results.

17. The published translation ('signpost') is inaccurate.

As explained in Chapter 4.2, Wittgenstein's rigorous mathematical verificationism was not only a corollary of the grammar view (since normativity cannot exist unbeknown), it was also suggested and corroborated to him by three other ideas of his early and middle period, namely: the postulate of the determinacy of sense, the postulate of complete analysability of a meaningful proposition (which in the mathematical case would result in a proof), and the demand that any search must be systematic (thus warranting a definite outcome). At least the second of those ideas—that in mathematics full analysis must involve proof—Wittgenstein continued to endorse in his later writings. Thus, he writes in 1944:

> Would one say that someone understood the proposition '563 + 437 = 1000' if he did not know how it can be proved? Can one deny that it is a sign of understanding a proposition, if a man knows how it could be proved?
>
> (*RFM* 295–6 [27.2.44])

A full understanding of a mathematical proposition (Wittgenstein suggests) would require us to know its full analysis, which amounts to a proof. A theorem and its proof stand in an internal relation to each other, such that the sense of the proposition is (at least partly) determined by its proof (*PG* 375; *RFM* 162b). Indeed, it appears tempting to identify the fully analysed mathematical proposition with its proof (*PR* 192), which also Wittgenstein continued to think, albeit less dogmatically than in 1930:

> Imagine that you have taught someone a technique of multiplying. He uses it in a language-game. In order not to have to keep on multiplying afresh, he writes the multiplication in an abbreviated form as an equation, and he uses this where he multiplied before.
>
> (*RFM* 296e)

In other words: the equation can be regarded as an abbreviation of the process of calculation.

For most of the time Wittgenstein doesn't distinguish between routine calculations and creative proofs, between 'homework' and mathematical research. But that is a distinction well worth emphasising and clarifying.

What about sums whose correctness we haven't checked yet: (if mathematical sense requires proof) would we not be forced to say that for all we know they may not be mathematical propositions at all, but lack sense?—That was not Wittgenstein's view in the early 1930s when he held that sums and homework questions are meaningful in virtue of the well-known method of calculation by which we check or answer them (*AL* 197–8). Thus, even an incorrect sum is meaningful insofar as it reflects an understandable 'mistake in calculation' (*AL* 200). We can understand it as a failed attempt to apply a well-known method.—Some of Wittgenstein's later remarks, however, point in a different direction (*RFM* I §§106–12: 76–9). Although it is true that our familiar techniques of calculation give a straightforward use and meaning to any sum of the form '$a \times b = c$', in the case of a mistaken sum there is no genuinely *mathematical* content.

A false sum, such as '*16 × 16 = 169*', is meaningful insofar as it can be applied outside mathematics (probably leading to an empirical error), and insofar as it can be checked by a calculation. However, if I mistakenly believe that *16 × 16 = 169*, I do not believe a mathematical proposition; rather, I mistakenly believe that '*16 × 16 = 169*' *is* a mathematical proposition (*RFM* I §111: 78). So, we should indeed say that an incorrect sum has no mathematical sense (i.e. does not have the status of a mathematical norm of representation), and hence, that an arithmetical equation we haven't checked yet may turn out to have no mathematical sense, even though it has a straightforward use: Someone mistaking it for a mathematical proposition will apply it accordingly, and someone sceptical about its correctness will know how to check it.

However, there is not only elementary school mathematics, the calculus of basic arithmetic everybody learns at school to use in everyday life, there is also mathematics as a discipline of research (whose results may later find an application in the sciences). Mathematical research is a matter of developing unknown implications of or extending existing calculi: developing new mathematical concepts that are (more or less) natural continuations of existing techniques. For instance, unrestricted use of subtraction leads to the introduction of negative numbers; unrestricted use of division leads to fractions. These are extensions of the calculus. In the original calculus of addition and subtraction with natural numbers, '5 – 7' is meaningless; like an illegal move in chess.

So, we need to distinguish *two types of mathematics*: *everyday mathematics*, which is *static*: calculations within a given calculus; and *mathematical research*, which is concerned with problems that cannot be solved by know algorithms within the existing calculus: the solution cannot simply be worked out using a given set of rules. Mathematical research is essentially *dynamic*: it does not just apply established algorithms in an existing calculus, but develops new ones or even extends the calculus.

In mathematical research, we ask questions of which we do not yet know how to find their answer, and we advance conjectures, would-be mathematical propositions, for which we do not have a proof, nor a method of producing a proof. But (as already seen in Chapter 4.2) with regard to mathematical research the claim that the meaning of a mathematical proposition is determined by its method of proof, appears to have three paradoxical implications. First, it would appear that —:

(P1) A mathematical conjecture has, for the time being, no determinate meaning.

At any rate, a conjecture cannot have the same meaning that it will have once a proof has been found. Hence, it would appear that

(P2) A conjecture can never be proven true.

For what is proven true must *ipso facto* be a different proposition from what was only conjectured (*PR* 183, 191; *RFM* 366d). Moreover, it would appear, very implausibly (*PR* 184b; *WVC* 109), that

(P3) There cannot be more than one proof for a given mathematical proposition.

Of course even in 1930 Wittgenstein wouldn't hold that mathematical conjectures are entirely meaningless; that they are just pieces of nonsense (*PR* 170). When wrestling with this problem in *Philosophical Remarks* and *The Big Typescript*, Wittgenstein came up with the following suggestions (already presented in Chapter 4.2):

(i) Even for new, difficult problems mathematicians already have a method of solving them, only not in writing, but in psychological symbols 'in their heads' (*PR* 176).
(ii) Mathematical conjectures and unsolved problems can be formulated only in mathematical 'prose', which is not part of mathematics proper (*PR* 188–9).
(iii) Mathematical conjectures have only a heuristic sense: directing mathematicians' thoughts in a way that may prove fruitful (*PR* 190).
(iv) Mathematical conjectures or research problems are not part of mathematics, but should be seen as concerns about how best to *extend* our existing mathematical system (*WVC* 35–6), comparable to the challenge of inventing a new game that fulfils certain requirements (*BT* 620).
(v) When we appear to understand a mathematical conjecture, we only understand a corresponding empirical claim (*BT* 617–9).

As discussed in Chapter 4.2, the first four of those suggestions are unsatisfactory. It remains to be seen then if Wittgenstein's later writings provide us with ideas for a more convincing account of mathematical research and a plausible solution to the three paradoxes (P1)–(P3).

I believe that is indeed possible. On Wittgenstein's considered view, it can be maintained that proofs contribute to the sense of mathematical propositions without denying that mathematical conjectures can have a reasonably clear and legitimate sense, too.

(a) *Proofs explain* how *a proposition is true*. Wittgenstein compares a mathematical proof to a jigsaw puzzle (MS 122, 49v). Indeed, sometimes he regards actual jigsaw puzzles as mathematical problems (*RFM* I §§42–50: 55–7, *LFM* 53–5). In such a case, the conjecture to begin with would be something like: 'These 200 pieces can be assembled to form a rectangular picture of a mountain'. Here it is obvious that the proof—putting all the pieces together in the right way—would do more than establish the truth of the conjecture. It would not only convince us *that* the pieces can be put together to form a picture of a mountain, it would show us *how* they fit together (cf. *RFM* 301d; 308b). Thus the proof does not only verify a proposition, one can say, at least in some cases, that it gives us a much fuller understanding of it, showing us what exactly that proposition means.

(b) *Proofs account for mathematical necessity*. Mathematical propositions are characterised by a necessity that must be established by a demonstration. If that is correct, then the proposition that there is no greatest prime should be rendered more appropriately as: 'There *can't* be a greatest prime'—indicating the necessity we attribute to a proposition when we take it as a piece of mathematics. Then, of course, the meaning of the modal verb in that sentence needs to be understood. One is entitled to ask: 'What do you mean by "can't"?' And the answer that gives meaning to the 'can't' is that it follows from such-and-such compelling considerations, spelling out the implications of our definitions—the proof—that there is no greatest prime.

In this way, the mathematical proposition, when taken as such: as a demonstrably necessary truth, refers us to its proof (cf. *RFM* 309).

(c) *Only proof shows a conjecture to be consistent and hence, ultimately, meaningful*. Consider that, for all we know, a mathematical conjecture could be proven false (*RFM* 314d), that is, shown up to be inconsistent. Yet if something is inconsistent, or contradictory, it does not make sense: it cannot be understood: there is nothing to be understood (*LFM* 179, 47; *RPP* II §290). But then, given that we cannot even know whether a mathematical conjecture is fully *understandable* (and not nonsense), then a fortiori we cannot claim to *understand* it. A sentence that as far as we know may be inconsistent, i.e. nonsense, can hardly be said to have a clear sense for us. This, again, vindicates Wittgenstein's view that a proof gives meaning to a mathematical proposition.

Or shall we say that it may well *have* a clear mathematical meaning all along, only we don't yet know it?—Well, as far as the contents of the proposition are concerned we could perhaps put it that way. (Wittgenstein recognizes that the expression 'sense of a mathematical proposition' is not sharply defined and can be construed in different ways (MS 122, 113).) After all, when at a later time a proof is found, it is accepted as a proof *of that conjecture* (e.g. Andrew Wiles's proof became famous as a proof of Fermat's Last Theorem). We take the proof to show that the conjecture had been correct, i.e. provable, and hence meaningful, *all along*.—However, another aspect of our concept of the sense of a linguistic expression is that it is the internal object of correct linguistic understanding. Normally, the sense of a linguistic expression is what a competent speaker understands by it. Hence, in a case where for the time being even the most competent speakers (mathematicians) do not fully understand a would-be mathematical proposition (nor know how to come by such an understanding), we can also with some propriety say that so far it has no clear sense for us (or anybody).

(d) *Proof affords normative legitimacy, which is a crucial part of the meaning of a mathematical proposition.* Given Wittgenstein's key ideas that mathematical propositions are akin to grammatical norms (*RFM* 162d, 169b, 199a, 320a), for a proposition to have mathematical sense it must not only have the contents, but also the *normative status* that characterises mathematics (*RFM* 425e): it must be acknowledged as a grammatical rule, which obviously an unproven conjecture is not (since we treat proof as a condition for according that normative status). Nothing unknown (or not known to be provable) can fulfil a normative function (*PR* 143; 176). Therefore, even if we assume that it is possible to find a proof for Goldbach's conjecture—that the potential for such a proof is already there—, until it has actually been produced Goldbach's conjecture will not be accorded the status of a grammatical rule. That is, until then it cannot legitimately be accorded the full status of a mathematical proposition.

Consider again the legal analogy mentioned earlier: A country has a complicated piecemeal system of laws, byelaws, statutes and regulations. Moreover, there are lawyers that have developed an equally complex system of rules as to how apparent conflicts between different pieces of legislation are to be resolved. One day a lawyer produces a legal proof that the President is to give the Annual Opening Speech in Parliament. Henceforth it is regarded as a *legal obligation* that the President is to give the Annual Opening Speech in Parliament. Of course, if the proof is correct, then it could have been given before. In a sense, the laws haven't changed; so implicitly they would have required the President to fulfil that duty all along. But then, as long as that implication wasn't acknowledged as such, it could not have had any normative force: it did not have the status of a law. Similarly, a mathematical conjecture, even if provable all along, must first be proven, before it can attain the status of a mathematical proposition.

When discussing Wittgenstein's comparison of a mathematical research question to the task of inventing a new game (*BT* 620; considered in Chapter 4.2 (ix)), I rejected the idea that solving a mathematical problem (or proving a mathematical conjecture) always involves introducing new symbols and concepts (e.g. signed integers). Even a proof that makes use only of existing concepts can be extremely difficult to find. And now, I hope, it can be seen that the change of meaning from a conjecture to a theorem[18] can be accounted for without the claim that a proof introduces new symbols and concepts. It follows from Wittgenstein's key idea that mathematical propositions are akin to grammatical rules and as such cannot exist unbeknown. Hence, even if (unlike Wittgenstein) we want to say that the connections developed by a proof were already implicit in the rules, they need to be drawn out and acknowledged before they can become part of our system of mathematics.

(e) *Mathematical conjectures (or problems) can have a fairly clear sense, but it's not a genuinely mathematical sense.* This idea is already sketched in *The Big Typescript* (see (iv)–(v) above). In as much as we understand the content of a mathematical conjecture, we take it as an *empirical* proposition, corresponding to, but crucially different from, the mathematical proposition we would like to establish by proof. For example, in some sense we understand the idea of the construction of a heptagon with ruler and compass (which is impossible). But that is only because we have a clear empirical idea of a heptagon, that is, we can easily think of a 7-sided figure whose sides and angles when measured come out as all the same. So we are inclined to understand the problem as that of drawing such a figure. But in fact that is not the mathematical problem. The mathematical problem is that of finding a mathematical *construction* of a heptagon, analogous to the way one can give a mathematical construction of, say, a pentagon. The result of such a construction would of course also fulfil the empirical criteria (that measurement shows 7 sides and angles, all roughly equal), but that is not enough. As a solution to a geometrical problem, it is essential that the figure be arrived at, step by step, in a regular, repeatable and teachable way, using only ruler and compass. We are looking not just for a shape, but for a very specific way of producing it. Yet this specific way of producing such a shape is something we are unable to describe. We have no idea of such a geometrical construction; and therefore, our

18. In that sense at least it can be said that every proof produces a change of meaning. It gives the proposition proven a different normative status.

talk of such a construction—the conjecture of such a construction—has no clear mathematical sense; even though it has a very straightforward empirical sense, derived from empirical measurements of drawn figures (cf. *PLP* 391–2).

As another example, consider Goldbach's conjecture that every even number is the sum of two primes. Don't we understand that?—Again, Wittgenstein's response is that without a proof we have of course some understanding of it, but only as an empirical generalisation; meaning that for any even numbers we will ever consider we will be able to find two primes adding up to that number. That is an empirical hypothesis inductively supported by our evidence to date; but not a mathematical proposition (cf. *RFM* 280–1).

Infinity in mathematics is always the endless applicability of a law (cf. *PR* 313–4; *RFM* 290b). Hence, where (as yet) we have no law, no mathematical rules that can be understood to have an endless applicability, we cannot meaningfully speak of mathematical infinity. So we cannot as yet make sense *of the infinite scope* of Goldbach's conjecture.

What we're looking for when trying to find a solution to a mathematical problem is a proof, which of course we cannot really describe before we've actually found it. For an accurate description of a proof is the proof itself. This brings out the difference between looking for some physical object and looking for something in mathematics: In the former case I can know exactly what I'm looking for. I can, for example, give you a perfectly accurate description of the spectacles I've mislaid. Not so in mathematics, where the thing looked for consists only in its accurate description (*PLP* 393). It follows that in mathematics we can never know *exactly* what we've been looking for till we have it. Hence a mathematical conjecture can never have a clear and precise mathematical sense, although, to be sure, we may be able to attach a fairly clear empirical sense to it, and some vague idea as to, roughly, what kind of proof strategy we expect might be successful.—

In conclusion, we need not attribute to Wittgenstein (P1) the paradoxical view that a mathematical conjecture has, for the time being, no clear meaning. Conjectures are surely meaningful; it is only that they are still waiting to be given a precise mathematical content and the normative status of a mathematical proposition. Does that mean that (P2) a conjecture can never be proven true: for what is proven true must *ipso facto* be a different proposition from what was only conjectured? Yes and no. In one important sense, what is proven is indeed a different proposition; but in another sense, it can obviously be recognised as delivering what the conjecture was asking for, namely that a given formula, of whose possible applications we may already have a fairly good understanding, be given the role of a theorem.

Finally, let us turn to the third paradox (P3), namely the claim that since the meaning of a mathematical proposition is determined by its proof, two different proofs will yield two different propositions. In other words: a mathematical proposition cannot have two different proofs.

Wittgenstein acknowledges that this is contrary to what we say in mathematics (*RFM* 189b). In fact it is quite common for one mathematical proposition to have more than one proof. For instance, at least six different proofs can be cited for the theorem that there is an infinity of primes (Aigner & Ziegler 2001, Ch. 1).

Let us note first that Wittgenstein's main reason for insisting on a difference in sense between a conjecture and the corresponding theorem does not apply in this case. The point is that without an endorsement by proof a conjecture cannot have the normative status of a mathematical proposition. By contrast, two different proofs will still agree in conferring this normative status on their conclusions. Consequently, where (what appears to be) the same mathematical proposition has been proven in two different ways, both proofs will confer on it the status of a norm of representation and thus qualify it to be used in a certain way, especially for applications outside mathematics:

> That these proofs prove the same proposition means, e.g.: both demonstrate it as a suitable instrument for the same purpose.
>
> And the purpose is an allusion to something outside mathematics.
> (*RFM* 367bc)

The question to what extent mathematical proofs are to be regarded as constitutive of the sense of their conclusions may be compared with analogous verificationist issues outside mathematics. Given that meaning is use, it is very plausible to insist on an important difference in meaning between, on the one hand, propositions that are susceptible of, and intended for, empirical confirmation or disconfirmation, and, on the other hand, propositions that do not allow, or are never intended to be exposed to, any empirical testing (e.g. articles of faith, moral principles). Thus Wittgenstein emphasised that the most prominent confusion in the philosophy of religion was due to neglecting this crucial distinction: treating expressions of religious belief as if they were cosmological hypotheses (*LC* 53ff.).

But Wittgenstein was inclined to go further. He would claim that even among the class of verifiable propositions differences in meaning resulted from different methods of verification. He held, for example, that statements of length had a different sense depending on whether they were to be established by using a tape measure or by astronomical calculations (*RFM* 146–7; *BT* 80v). Or again, the meaning of a proposition about the time of death would partly depend on the methods by which the time of death is ascertained (MS 122, 113v–114r). This further claim is, of course, very convincing where different methods introduce different criteria that can lead to different results in the same case. Thus, depending on whether a diagnosis of death is based on cessation of heart action or on failure of brain activity, the proposition that the patient died at 5.30 p.m. may indeed have different meanings.

It is, however, doubtful whether this line of reasoning can ever be applied to mathematics, where different and inconsistent definitions would hardly be tolerated. Thus, different proofs of the infinity of primes would surely not involve different and inconsistent definitions of a prime number.

Indeed even with regard to a mathematical proposition's role inside mathematics, some of Wittgenstein's later remarks suggest that different proofs need not be seen to give the proposition different meanings. The idea now is that a proof alone is anyway not *sufficient* to endow a mathematical proposition with meaning. Rather, its meaning depends on a whole system of rules and techniques, only some of which are made explicit in the proof:

> The proof of a proposition certainly does not mention, certainly does not describe, the whole system of calculation that stands behind the proposition and gives it its sense.
>
> (*RFM* 313d; cf. 367d)

A proof locates a proposition in our system of mathematics; and there may be different proofs, that is, different ways of reaching the same position. Yet the proposition's sense would be its position in the system, not the path by which it is reached:

> A proof of the proposition [that there are infinitely many prime numbers] locates it in the whole system of calculations. And its position therein can now be described in more than one way, as of course the whole complicated system in its background is presupposed.
>
> (*RFM* 313b)

Nonetheless, a proof can be said to *determine* the sense of a proposition, since it determines its location in the system. Moreover, every new proof enriches our system of mathematics, by adding new conceptual links to it. Hence it could still be maintained that by finding a new (second or third) proof for an already established theorem we add to its meaning, or at any rate to our understanding of it. (That is especially plausible when after a non-constructive proof, a constructive one is found (*RFM* 282d; 308b).)

In fine, Wittgenstein's considered view does not commit him to (P3): on balance he would not insist that the same mathematical proposition cannot have more than one proof.

10.3 What Is the Relation Between a Mathematical Proposition's Proof and Its Application?

Wittgenstein's 1929 idea that mathematical propositions are given sense by their proof can be seen as just a special case of a comprehensive verificationism: the view that the meaning of a proposition is the method of its verification (*PR* 200; Frascolla 1994, 58–63). Yet, as noted before, there is a crucial difference between the empirical and the mathematical realm. In the former we may know of a method of verification (or falsification) in principle, without being able to apply it in a given case. We may, for example, describe the kind of evidence that would settle a given historical hypothesis without being able, as it happens, to get hold of such evidence. In mathematics, by contrast, once we *know* a method of verification we can apply it, since mathematical methods of verification don't rely on fortuitous contingencies (apart from such trivial requirements as the availability of pen and paper). A detailed *description* of what would verify a mathematical proposition does already amount to its verification: 'the mathematical proof couldn't be described before it had been discovered' (*BT* 631). Therefore, in mathematics for the verificationist requirement of *verifiability* to be fulfilled we must already be in a position to provide an actual *verification*.[19]

However, during the 1930s Wittgenstein realised that the method of verification was only one aspect of the sense of a proposition and, moreover, that not all meaningful declarative utterances were verifiable. This development went together with a fundamental change in Wittgenstein's view of language: from the calculus model of the *Tractatus* to the language-game view of his later philosophy (Baker & Hacker 1980, 89–98; Glock 1996, 67–72; Schroeder 2013, 155–9). In the *Tractatus* Wittgenstein offered a mathematical picture of language: presenting language as a calculus. The essence of language, the general form of the proposition was given by a simple formula (*TLP* 6). Like a calculus, language was claimed to be governed by syntactic rules: (i) formation rules about the licit combination of names to form elementary propositions; (ii) formation rules about the licit combination of elementary propositions to make complex propositions; and finally (iii), truth-table rules, which enable us to identify logical truths and entailments. In 1930 Wittgenstein still claimed that 'Language is a calculus' (*LL* 117), but in the following years he gradually abandoned the calculus model. He came to

19. As explained (in Chapter 4.2), another consideration that led Wittgenstein in 1929 to the claim that the meaning of a mathematical proposition is specified by its proof was his *Tractatus* idea of complete analysability (*TLP* 3.25). If we think of an ordinary arithmetical equation, such as '*25 × 25 = 625*', it is clear that its complete analysis will *be* its verification (cf. *PR* 179).

realise that language is different from a logical calculus in the following respects:

(i) Linguistic competence, unlike the mastery of a logical or mathematical calculus, is not based on the learning of rules (cf. *PI* §54; Schroeder 2013, 157–62). And this is an important point for Wittgenstein, for he identifies as a typical philosophers' mistake an inclination to describe our learning of language in a way that already presupposes linguistic competence, like the learning of a foreign language (*PI* §32)—or indeed the learning of a mathematical calculus. Therefore he emphasises that at the basic stages language teaching cannot be explaining, let alone giving rules, but must be training or drill ('*Abrichtung*') (*PI* §6).[20] And even at later stages, language is for the most part learnt by picking up a sufficient number of examples of correct usage, rather than by rules or explicit explanations. 'One learns the game by watching how others play' (*PI* §54). Whereas the very idea of a calculus is that every transformation is carried out according to explicitly stated rules, natural languages are used long before linguists give summarizing descriptions of that use in terms of grammatical rules (*BT* 196, *PI* §54). Therefore, Wittgenstein acknowledges in the early 1930s that language is not literally a calculus, it can only be *compared* to a calculus (*BT* 196, *BB* 25, *PI* §81, §131).[21]

(ii) Wittgenstein observes that, unlike the operation of a calculus, the use of language is not everywhere bounded by rules (*PI* §84): there is frequently vagueness (*PI* §88) and indeterminacy (*PI* §80). The meanings of most words are not given by definitions, but by examples, and it may not even be possible to find a definition retrospectively, as concepts can also be based on an irregular network of 'family resemblances' (*PI* §66).

(iii) Most importantly, Wittgenstein stresses that speaking a language is not just uttering or manipulating signs, but that our use of language is essentially embedded in our lives, interwoven with all sorts of non-linguistic behaviour. For this ensemble of uttering signs in connection with various other kinds of behaviour (e.g. the verbal and non-verbal transactions between customer and shopkeeper (*PI* §1)) Wittgenstein introduces the term 'language game' (*PI* §7). While a calculus may be an appropriate model or object of comparison for the transformations

20. Cf. MS 179, 7v: 'To follow a rule presupposes a language.' (*Einer Regel folgen setzt eine Sprache voraus.*)

21. Note that Wittgenstein did not for that matter retract his key idea that language is essentially normative. It is only that he came to realise that linguistic norms, the distinctions between correct and incorrect uses, are, for the most part, picked up by piecemeal exemplification and emulation, rather than by learning rules, i.e. generalising verbal instructions (see Schroeder 2017b, 258–60).

of one linguistic expression into another—say, when an inference is drawn—the concept of a language game highlights the connections between linguistic utterances and other human actions—for example, when an order is carried out. That is the later Wittgenstein's anthropological perspective on language: for a thorough understanding of linguistic phenomena it is not enough to study the relations among words and sentences; instead, we need to see linguistic utterances as human actions and words and sentences as tools for such actions (*PI* §569). The meaning of a word is its use (*PI* §43), a sentence is an instrument and its sense is its employment (*PI* §421).

Wittgenstein's earlier verificationist claims were not abandoned, but restricted to only *some* uses of language and integrated into his broader anthropological view of language. It is indeed an essential characteristic of the language games of making empirical claims that they are understood to be subject to confirmation or disconfirmation: that we know how their truth or falsity could affect our experience. But then, of course, there are other language games—such as expressing one's preferences, poetry, or religious rituals—that lack such a verificationist element.

In Wittgenstein's thoughts on mathematics we find similar developments. After a verificationist phase around 1930, presenting proof as the source of mathematical meaning, the later Wittgenstein focuses more on use. On the 28th October 1942 he emphasises that for a calculus and its demonstrations to qualify as mathematics they must have an application:

> I want to say: it is essential to mathematics that its signs are also employed in *mufti*.
>
> It is the use outside mathematics, and so the *meaning* of the signs, that makes the sign-game into mathematics.
>
> (*RFM* 257de)

And yet he holds on to the calculus view with the corollary that the meaning of a theorem is due to proof. Little more than a fortnight later, on the 15th November 1942 he writes:

> In a certain sense it is not possible to appeal to the meaning of the signs in mathematics, just because it is only mathematics that gives them their meaning.
>
> (*RFM* 274d)

And again (in February 1944):

> The problem of finding a mathematical decision of a theorem might with some justice be called the problem of giving mathematical sense to a formula.
>
> (*RFM* 296b)

Wittgenstein was not unaware of the tension between those two key ideas. He concedes that his old verificationist slogan that proof determines meaning needs to be qualified. There is truth in it, but also falsehood (*RFM* 367d [16.6.41]): 'it is as if the proof did not determine the sense of the proposition proved; and yet as if it did determine it' (*RFM* 312g).

Undoubtedly, proof is *a* source of mathematical meaning. Somebody who believed the proposition '*563 + 437 = 1000*' on authority alone, unable to work it out himself, could hardly be credited with a *full* understanding of it, whereas someone who could demonstrate that *563 + 437* equals *1000* would thereby manifest at least some understanding of it (*RFM* 295–6). And even though, obviously, there can be some understanding of mathematical conjectures, once we are given a proof—say, of Fermat's proposition—we understand it better than before (MS 127, 161).

So proof is a determination of meaning, but it is not the only one. After all, 'one can know a proof thoroughly and follow it step by step, and yet at the same time not *understand* what it was that was proved' (*RFM* 282f). That is partly because a proof is only one instance of a complicated and open-ended 'system of calculations that stands behind the proposition and gives it its sense', without being made explicit in the proof (*RFM* 313d). But even somebody who has mastered a whole system of proving theorems—e.g. in Russell's *Principia*—may calculate only mechanically, without understanding the meanings of the signs, and so not have a full grasp of the propositions proven (*RFM* 257–8).

Moreover, the application of mathematical propositions is separable from their proof, as the following thought experiment illustrates:

> Imagine that calculating machines occurred in nature, but that people could not pierce their cases. And now suppose that these people use these appliances, say as we use calculation, though of that they know nothing. Thus e.g. they make predictions with the aid of calculating machines, but for them manipulating these queer objects is experimenting.
>
> (*RFM* 258ef)

These people use arithmetical equations as quasi-grammatical propositions, just as we do, although they cannot work them out, but *find* them. (This is perhaps not so different from the way some mathematically clueless school children derive results from a calculator.) In this case, we may assume that there *are* proofs for the propositions in question, even though they are not accessible to people and play no role for them. But, as Wittgenstein notes in 1944, 'one can also apply an unproved mathematical proposition; even a false one' (*RFM* 435f).

Thus we have to acknowledge *two* sources or aspects of mathematical meaning: proof and application. Occasionally Wittgenstein suggested that they can be brought together: that proof, or at least a proper understanding of a proof, would encompass an understanding of a theorem's applicability:

> The proof, like the application, lies in the background of the proposition. And it hangs together with the application.
>
> (*RFM* 304e)

> When the proposition seems not to be right in application, the proof must surely shew me why and how it *must* be right; that is, *how* I must reconcile it with experience.
>
> Thus the proof is a blue-print for the employment of the rule.
>
> How does the proof justify the rule?—It shews how, and therefore why, the rule can be used.
>
> (*RFM* 305cde)

> The proof shews how one proceeds according to the rule without a hitch.
>
> (*RFM* 306g)

In what way could the proof of a theorem guide, or be a proof of, its applicability?

As argued above, proofs spell out logical implications of the definitions of mathematical concepts. In some cases it may be said that a mathematical concept was, to begin with, underdetermined and the proof produced a further determination: developing the concept in one way, rather than another. That is Wittgenstein's idea of mathematical proof as concept change, which we found may be correct in some cases, but not in others. Be that as it may, in either case, a proof can be said to draw out the implications of certain mathematical concepts. The question would only be whether all of the relevant concepts had been sharply defined all along or whether some of them were only sharpened up, implicitly, by the proof itself. In other words, taking a proof to be of the form:

> (P4) From the concepts Φ and X it follows that what is φ and χ must also be ψ,

—on Wittgenstein's model the conclusion would imply a definitional addition to the initial concepts Φ and X. Thus, a more explicit schema of the proof would be:

> (P5) From the concepts Φ and X, *understood to be defined to have feature* Ω, it follows that what is φ and χ must also be ψ.

If that is an appropriate conception of a mathematical proof, then a theorem's applicability can only depend on the applicability of its concepts (as taken or construed in the proof). Consider again Wittgenstein's example of overlap addition (*RFM* I §38: 52). A proof that *8 † 8 = 15* could certainly show us 'how it must be right'—by clarifying the meaning of the operation sign '†'. That is, *if* we find occasion to apply this operation the proof convinces us that it must be possible to apply the theorem '*8 † 8 = 15*'. But of course, there is no guarantee that any useful application can be found. A mathematical proof (especially at a higher level) can certainly not direct us to an *empirical* application of its results.

As already noted earlier, it would be unrealistic to expect every piece of mathematics, every theorem to find an application outside mathematics. Nor does Wittgenstein make such a demand. All he is inclined to insist on is that mathematical *signs*, or concepts, are also used outside mathematics (*RFM* 257e). But even that may not be true in all cases. For example, it has been proven (by Euclid) that for every Mersenne prime number m ($= 2^p - 1$, where p is itself a prime) there is a perfect number P (i.e. a number that is the sum of its proper factors), namely: $P = 2^{p-1}(2^p - 1)$. For that to be a respectable mathematical theorem it is surely not required that the concepts of a Mersenne number and a perfect number (or their signs) find an employment outside mathematics. Even if the basic concepts Φ and X (in our schema of a mathematical proof)—say, the concepts of natural numbers an arithmetical operations—have an employment in mufti, that need not be true of all the more abstract concepts that can be defined in terms of Φ and X—e.g. Mersenne prime—and then occur in theorems.

In fine, it sounds plausible to insist that it is essential for our mathematical practice to be conceptually connected with empirical language, but that can hardly be a requirement for every individual mathematical proposition or concept. And which of our mathematical concepts and theorems find an empirical application can surely not be read off a proof.

Nonetheless, it may be possible to bring together proof and application, if as a starting point we take a sufficiently broad concept of use.

> The use of the signs must settle [the sense of a proposition]; but what do we count as the use?—
>
> (*RFM* 366e–7a)

The way a certain kind of linguistic expression or sentence is used Wittgenstein also calls a language game (*PI* §23), yet a language game includes both what *leads to* the formulation of a certain utterance and what *follows it*. Thus, the use or language game of a scientific proposition includes both the way it has been derived from repeated experiments, hypothesising, and verification, and the way it can then be applied, say, in engineering. Similarly, the mathematical language game includes both proofs and—wherever possible—applications.

When from his verificationist phase in the 1930s Wittgenstein moved on to a conception of meaning as use, the idea that the sense of a proposition was given by its method of verification was not rejected as false, but qualified as only a partial answer. Wittgenstein held on to the view that an explanation of how a proposition can be verified 'is a contribution to the grammar of the proposition' (*PI* §353); only it is not a complete account. Meaning is use, or the role of a linguistic expression in a language game, which in the case of empirical propositions *includes* methods of verification. Likewise, in the philosophy of mathematics, Wittgenstein moved from the idea that the sense of a mathematical proposition was determined by its proof to a more comprehensive account of meaning as use, including both proofs and applications.

Finally, we should reconsider the issue of mathematics' prognostic usefulness, first raised in Chapter 6.

On the one hand, Wittgenstein insists that mathematics only presents a framework for descriptions (*RFM* 356f) and can tell us nothing *about the future*. If I put first 5 apples on the table and then another 7, arithmetic cannot tell me what a *subsequent* count of the total will yield—that depends on the physical properties of apples (*PLP* 51–2); it can only provide a redescription of how many apples I *have* put on the table, namely: that '5 and 7' can also be called '12'.

On the other hand, we do use mathematics in order to *predict* what will be the case (*RFM* 232b, 356a), for example, how many tiles will be required to cover a certain surface area. Is mathematics' undeniable prognostic usefulness not incompatible with the grammar view (MS 163, 62r)?

The point to remember is that every *application* of mathematics leads to an *empirical* claim about the result of a count or a measurement. The mathematical transformation itself is a priori, but because it results in an expression with different criteria of application it corresponds to (the prediction of) a different empirical observation. That if you have 25 bags of 25 nuts each, you *have* 625 nuts in total is analytic; but that you *count* 625 nuts is empirical. The empirical (predictive) element comes in not only with regards to *future* observations, but already with counts or measurements at the same time.

In $_{N+}$English 'I have 625 nuts' is just a paraphrase of 'I have 25 times 25 nuts', but in $_{N}$English, which is the English we use simply to describe our counts (to which we temporarily revert when we haven't done a calculation yet or when we count to check a calculation), it is a distinct claim that we predict will be found true, too.

Ultimately, what the prediction amounts to is that our mathematical concepts will continue to fit our ordinary observational concepts. For example, that the arithmetical operation of addition corresponds to the physical operation of adding objects to other objects, and that our arithmetical concepts of natural numbers corresponds to the number concepts embodied in transitive counting.

One may be inclined to object: 'But surely, even if objects vanished when put together or recounted, it would still be true that in having 25 times 25 nuts I had, *ipso facto*, 625 nuts.' And that is of course true, but in saying so we are using ${}_{N+}$English, whose empirical applicability is not guaranteed by maths. It is imaginable that (reverting to ${}_{N}$English) in such a case I count 25 times 25, but not 625 (*RFM* 358f). Perhaps I regularly get confused when counting with three-digit numbers. The prognostic element lies in the assumption that the concepts of ${}_{N+}$English fit our experiences of counting objects (and will continue to do so).

In terms of the three apparent sources of mathematical truth (considered in Chapter 9 above), the situation could be described as follows: We have to choose our conceptual *conventions* in such a way that drawing out (or sometimes extending) their implications in calculations or *proofs* their applications are found *empirically* true.

11 Inconsistency

Hilbert's Proof Theory. One of the major players in the discussion about the foundations of mathematics during the first decades of the 20th century was David Hilbert with his programme of proof theory (also called Formalism). Like Russell, Hilbert was concerned to show that mathematics was absolutely reliable, a concern partly fuelled by the set-theoretical paradoxes that had upset Frege's foundationalist project and also by an apprehension that our familiar logic (with the Law of Excluded Middle) might run into trouble when applied to infinite totalities, as invoked in analysis (in the form of irrational numbers) and further developed in Cantor's theory of transfinite numbers. Unlike intuitionists, Hilbert was not willing to settle for a reformed body of mathematics, pruned of its liberal engagement with infinite totalities. He famously declared that: 'No one shall drive us out of the paradise which Cantor has created for us' (Hilbert 1925, 141). Against the logicist project he objected that some of its axioms were implausibly arbitrary and, in any case, the ultimate subject matter of mathematics was prior to the application of logical operations:

> something must already be given in representation [*Vorstellung*]: certain extra-logical discrete objects, which exist intuitively as immediate experience before all thought.
>
> (Hilbert 1922, 202)

Hence, the theory that Hilbert put forward in the 1920s 'to establish once and for all the certitude of mathematical methods' (1925, 135) was as follows. First, arithmetic was to be formalized: reconstructed as an axiomatic system. For that purpose, Hilbert borrowed approvingly from the achievements of Frege and Russell, but unlike the logicists he did not propose to produce a calculus constructed entirely in terms of second-order logic and set theory. Rather, axioms of logic were to be combined with genuinely mathematical axioms in terms of numerical symbols (initially in stroke notation) and symbols for arithmetical operations (Mancosu 1998, 129). In order to cover all of classical mathematics, Hilbert's axiomatic system was also to contain postulations of 'ideal elements' (comparable to the point at infinity where parallels are said to meet in projective geometry), such as transfinite cardinal numbers. However, this enriched axiomatic system with its ideal elements is to be regarded in an instrumentalist, or game-formalist, way (Mancosu 1998, 136). Although it is grounded in experience (our intuitions of small natural numbers) and its results may again be applicable to the empirical world (in trade, engineering, or physics), individual expressions inside the calculus need not be taken to refer to anything:

> To make it a universal requirement that each individual formula is interpretable by itself is by no means reasonable; on the contrary, a theory by its very nature is such that we do not need to fall back on intuition in the midst of some argument. What the physicist demands of a theory is that propositions be derived from laws . . . solely by inference, hence on the basis of a pure formula game, without extraneous considerations. Only certain consequences of the physical laws can be checked by experiment—just as in my proof theory only the real propositions are directly capable of verification.
>
> (Hilbert 1928, 475)

But then, as mathematical results, unlike physical theories, are not tested empirically, by trial and error, what is the legitimacy of such a calculus with 'ideal elements'? Hilbert's answer is: a proof of consistency which, in order to avoid circularity, does not use any non-finitary means (such as induction over infinite domains). This second part of his programme is called meta-mathematics. It is to be understood as a second-order theory about the axiomatic system that is mathematics proper. While the latter contains only 'purely formal modes of inference', in meta-mathematics 'we apply contentual inference' (Hilbert 1922, 212), yet the content matter is just the signs of our axiomatic system, finite combinations of symbols.

Having read Hilbert in 1930, Wittgenstein first of all objected to Hilbert's conception of meta-mathematics as a theory about mathematics proper. Rather, 'meta-mathematics' relates to ordinary mathematics as the 'theory of chess' relates to chess. Yet:

> What is known as the 'theory of chess' isn't a theory describing something, it's a kind of geometry. It is of course in its turn a calculus and not a theory.
>
> (*WVC* 133)

What is Wittgenstein's concern in denying such meta-discourse the status of a theory? A *theory* is taken here as a systematic description of an independently existing subject matter, whereas formulae in a *calculus* are correct not because they correctly describe something, but simply because they have been derived according to the rules (*WVC* 141). It is of course essential to Wittgenstein's overall view that mathematics should not be regarded as a theory (say, describing some Platonic objects), but as a network of calculi with empirical applications. And Hilbert's meta-mathematics for him is just such another mathematical calculus (*WVC* 121). That is because the criteria of correctness are not those of a correct observation, but simply applications of rules. Analogously, if 'chess theory' has it that a certain endgame position cannot be won, that is not based on empirical studies of actual games, but shown straightforwardly by considering the rules of chess when applied to that position. Again, algebra is not a theory generalising observations of arithmetic, but another calculus based on arithmetical rules applied not only to numbers, but also to variables (*WVC* 136). Likewise, Hilbert's meta-mathematics is a more advanced mathematical calculus, rather than a quasi-empirical theory *about* mathematics (cf. Mühlhölzer 2012).[1]

More importantly, Wittgenstein insisted that there was no need to find a consistency proof. In 1930 he remarked provocatively:

> I've been reading a work by Hilbert on consistency. It strikes me that this whole question has been put wrongly. I should like to ask: *Can* mathematics be inconsistent at all?
>
> (*WVC* 119)

What he means is that it is not obvious how a calculus can be said to be inconsistent.

According to Wittgenstein, a calculus as such can only be inconsistent by having contradictory rules, such that in a certain situation we are told to do something and also not to do it. Then we are stymied: we just don't know what to do. However, such a problem can easily be resolved. We simply qualify one of the conflicting rules to allow for an exception in line with the other one (*WVC* 120).

What Hilbert has in mind is something different. It is the contradiction between two assertions. In 1930 Wittgenstein seems inclined to regard that as strictly speaking not a problem within mathematics, but arising only in 'prose' (*WVC* 120), and to be avoided by a more careful analysis of what actually happens in the calculus (*WVC* 122). Later, however, he came to stress that for a calculus to be mathematics it must be more than a mere calculus: its concepts need to have empirical applications: 'It is the use outside mathematics . . . that makes the sign-game into mathematics'

1. Simon Friederich (2011) argues that beneath the obvious differences between Hilbert and Wittgenstein there is a fundamental agreement in that Hilbert's axiomatic method, too, amounts to a non-descriptive, normative account of mathematics.

(*RFM* 257e [1942]). Thus, it is essential to arithmetic that number signs are used to count and quantify objects. And while in an uninterpreted calculus the configuration '*p* . ~*p*' may be shrugged off with equanimity, deriving contradictory equations in arithmetic would be a different matter. For an equation such as '*7 + 5 = 12*' we do regard as something like a grammatical proposition, which we cannot easily hold on to together with its negation.

As discussed earlier, Wittgenstein was indeed inclined to regard arithmetical equations as akin to grammatical rules. But as such they are not just rules in a calculus (defining a game of writing down symbols), but rules about how to handle quantities, from which when applied in a given situation empirical propositions follow. For example, having first put 7 apples into a basket and then another 5, it follows that 12 apples have been put into the basket. Hence, in a given situation arithmetic instructs us to make an empirical assertion, e.g.: '12 apples have been put into the basket'. If we could also derive the equation '*7 + 5 = 15*', we would be led to a contrary empirical statement, namely: '15 apples have been put into the basket'. We would have an inconsistency between empirical assertions.

So, given Wittgenstein's account of the status of arithmetic as essentially linked with empirical discourse, he can no longer dismiss concerns about consistency as easily as in 1930 by saying that a calculus makes no assertions and therefore cannot be inconsistent.

In his 1939 lectures and his 1940 manuscript remarks, Wittgenstein is still equally dismissive of Hilbert's and others' concerns to make sure that mathematics does not contain any hidden inconsistencies, but his arguments have developed. His provocative position is that contradictions are harmless. We can ignore or rectify them when they arise. There is no such thing as a 'hidden contradiction', for as long as a contradiction hasn't surfaced and we don't know how to derive it, it does not exist (*LFM* 209–29, *RFM* 202–21).

His position can be broken down into four claims:

(i) Inconsistencies are due to carelessness in our stipulation of rules (*LFM* 213, 221–2; *RFM* 202–4; *BT* 549). For example, the statutes of a country may lay down where on feast days the vice president has to sit, and it turns out that there is an inconsistency in the rules so that under certain circumstances we cannot follow them all (*LFM* 210). In arithmetic, suppose that second-order exponentiation has been introduced without concern about scope (*LFM* 218). We just write, say: 2^{3^2}. Sometimes we calculate $(2^3)^2 = 8^2 = 64$, but sometimes we work it out like this: $2^{(3^2)} = 2^9 = 572$. In other words, we didn't realise that the order in which you carry out the two operations involved makes a difference, and so we neglected to indicate that order in our notation. The result is an ambiguous symbol.

(ii) When eventually such inconsistencies come to light the problem is easily resolved: by introducing a new rule or by restricting one of the conflicting rules. 'When a contradiction appears, then there is time to eliminate it' (*LFM* 210). For it would only be a local problem, not affecting other, more straightforward uses of the calculus. Thus, the contradiction found in Frege's logic did not invalidate any of the valid arguments that the calculus could be used to prove (*LFM* 227; Marion & Okada 2013, 70).

(iii) A logician's concern is that from an inconsistency (such as two contrary results) one can derive an outright contradiction (of the form '*p* . ~*p*'), from which in turn, in formal logic, one can derive just anything: *ex contradictione (sequitur) quodlibet*, the Principle of Explosion. Wittgenstein's response to this concern is: 'Well then, just don't draw any conclusions from a contradiction' (*LFM* 220).

In other words, Wittgenstein alerts us to an important distinction. The fact that it is possible to derive a contradiction in a calculus, and from it just anything else, doesn't mean that it's possible for that to happen *inadvertently*. Even if it was possible in accordance with the rules to derive all sorts of nonsense, we wouldn't want to do it and so, as it's not the sort of thing one could do without noticing, we wouldn't do it. Hence, the mathematical demand that it shouldn't even be possible is more a form of purism than of any practical importance (*RFM* 371–2).

(iv) Mathematicians such as Hilbert are worried about possible hidden contradictions. Although arithmetic appears to be a consistent calculus, how can we be certain that no inconsistency will come to light in future? Do we not need a formal proof of inconsistency to rule that out?—No, replies Wittgenstein, the very idea of a 'hidden inconsistency' is misguided. Possibilities we're not aware of are not, as yet, part of our calculus. And in mathematics you cannot even look for something if you don't know how.

Among the students to whom Wittgenstein presented his view on inconsistency in his 1939 lectures was Alan Turing, who was not convinced. He raised two critical replies to Wittgenstein's ideas:

First, in response to (iii), the Principle of Explosion can make itself felt even where a contradiction has never been made explicit: one can 'get any conclusion which one liked without actually going through the contradiction' (*LFM* 220). Even if '*p*' and '~*p*' never occur in the same line, any conclusion '*q*' can be derived. So the damage can occur even if we are careful never to use a premise of the form '*p* . ~*p*'. This objection was presented again and more emphatically by Charles S. Chihara who scathingly suspected 'that Wittgenstein failed to grasp it' (Chihara 1977, 330).

Secondly, in response to (iv), Turing urges that the damage can be quite serious, suggesting that the use of a calculus with a hidden contradiction may lead to accidents in engineering, such as the collapse of a bridge (*LFM* 218).

On *Turing's first objection*:

(1) It is true that, for the deriving of arbitrary conclusions, '*p*' and '~*p*' need not occur as a conjunction, but they both need to appear as premises in a fairly conspicuous way. In other words, although the Principle of Explosion doesn't require inferences from '*p* . ~*p*', it does require inferences from '*p*' *and* '~*p*', which is not something we might easily do without noticing (Shanker 1987, 242–3). It is certainly not unreasonable or unrealistic to set oneself the aim not to derive anything from both a formula and its negation. To then end up doing so anyway would either betoken carelessness or confusion, that is, an objectionable lack of familiarity with our calculus and procedures.

When Chihara considers Wittgenstein's response to Turing's point he quotes him saying: 'The only point would be: how to avoid *going through* the contradiction unawares' (*LFM* 227). If anything, this suggests the opposite of Chihara's scornful suspicion. For here Wittgenstein acknowledges that it is possible not to be aware of a contradiction, presumably because it can be split up, with its parts '*p*' and '~*p*' occurring in different lines. So with respect to the Principle of Explosion, our only—and fairly trivial—concern would have to be to pay sufficient attention to our calculus not to overlook the occurrence of both a formula and its negation as premises.

In fact, it is clear that the 'explosion' would be quite obvious and easy to spot anyway as it involves the peculiar step of writing down an entirely *arbitrary* formula. No logician that was moderately awake could fail to notice when at some point he writes down *whatever formula he likes* (e.g. by the rule of disjunction introduction) and then, in the end, retains that arbitrarily chosen formula as a conclusion.

(2) Concern about the Principle of Explosion is in any case an instance of what Wittgenstein called the '"the disastrous invasion" of mathematics by logic' (*RFM* 281f). In arithmetic no such rule exists. So if an inconsistency was found in arithmetic, we would not be licensed within our arithmetical calculus to infer from it whatever we liked. To take our earlier example of an ambiguity in exponentiation again, if getting different results for 2^{3^2} we then wrote out the contradiction: '$64 = 572$ and $64 \neq 572$', no rule of mathematics would allow us to continue: 'therefore $2 \times 2 = 500{,}001$'. As Wittgenstein notes: 'we should not call this "multiplication"' (*LFM* 218). In such a case, formal logic does not do what it is supposed to do, namely give a formalised and more precise account of rational procedures in mathematics. Rather, it introduces an utterly bizarre inference rule that no sensible mathematician would ever

be inclined to apply. On encountering an inconsistency—say, calculating that a certain value $x = 180$, when a little earlier we found that $x = 181$—only a madman would infer that 'therefore $2 \times 2 = 500{,}001$'.

Turing does not distinguish between mathematics and logic. He speaks of 'the case where you have a logical system, a system of calculations, which you use in order to build bridges' (*LFM* 212). In the same vein, Chihara speaks of engineers and scientists using a 'logical system' (Chihara 1977, 334–5). As a matter of fact, they don't. Engineers apply mathematics, but not formal logic. There is a crucial difference between mathematics and logic. Mathematics is an extension of our language, giving us new concepts and operations in order to deal with quantities and shapes. It needs to be learnt and applied on top of ordinary language. Formal logic, by contrast, is not an extension of ordinary language, but just a reconstruction or modelling of some of its features: its inherent logic. Formal logic aims to give a neater, artificially precise, redescription, or analysis, of the forms of inference we carry out anyway in ordinary language or mathematics. Unlike mathematics, formal logic does not substantially enrich our conceptual repertoire. That is why, unlike maths, it is not taught at school. It has no practical application. In order to solve a quadratic equation, you first need to learn a new method or formula. In order to reason logically, you don't need to attend logic classes: they will only teach you labels for forms of inference you've been familiar with all along.

But once we remind ourselves that engineers don't apply calculi of formal logic, it becomes obvious that the logician's fear of a contradiction '$p\ .\ {\sim}p$' triggering the Principle of Explosion is quite unrealistic. The mathematical calculi used by engineers may lead to inconsistent results, but engineers would probably not derive a proposition and its negation. And even if they did, they would not apply a rule of *formal* logic, such as the Principle of Explosion, to it.

(3) In any case, Chihara is quite mistaken to think that blocking inferences from '$p\ .\ {\sim}p$' was Wittgenstein's key idea in response to inconsistencies. It was presented only as one possibility, as Wittgenstein was fully aware that inconsistencies could make themselves felt in other ways. Should things go wrong because we draw inferences from a contradiction, Wittgenstein recommends that we simply refrain from doing so. But he also considers inconsistencies that do not involve a formal contradiction.[2] An example he mentions repeatedly is that of division by zero: from $4 \times 0 = 5 \times 0$ you can if you allow division by 0 derive that $4 = 5$ (*LFM* 221). Thus, any false equation could be proved, without ever invoking the Principle of Explosion or even producing a formal contradiction.

2. I was reminded of this point by Harry Tappenden.

Wittgenstein's point is simply that we identify the source of such inconsistent results—here division by *0*, in the example above an ambiguity in second-order exponentiation, or indeed conclusions drawn from a contradiction—and remove it. We need to be sufficiently familiar with our calculus to understand how, in a given case, an inconsistency can arise, and then avoid the paths that lead to it (*RFM* 209c).

(4) An analogy with ordinary language is illuminating. The Liar Paradox can be taken to show that an ordinary language such as English is inconsistent in the following sense: using ordinary semantic rules, especially concerning the use of the word 'true', we can construct a proposition that is true if and only if it is false. Assuming that it must be true or false we end up with a contradiction. Does that mean that English is a defective language and arguments in English (say, in a scientific study) cannot be trusted? Of course not. All it means is that a certain very peculiar self-referential use of the truth predicate is to be avoided. We are obviously not to construct a Liar-type contradiction and then, by the Principle of Explosion, derive arbitrary empirical assertions from it (*RFM* 376h, *LFM* 213).

Chihara is outraged by Wittgenstein's quick dismissal of the Liar Paradox, which, he protests, has perplexed 'some of the best minds in logic and philosophy' (Chihara 1977, 335). He obviously fails to grasp Wittgenstein's point. It is not to deny that one may well want to find an analysis that dissolves the puzzle; it is just that even if it cannot be resolved: if it really does bring out an inconsistency in our system of semantic rules, that has no tendency to taint or undermine other (and more important) uses of the concepts involved. Wittgenstein wants us to see inconsistency from an anthropological perspective (*RFM* 220b; cf. *WVC* 201). We are to understand that *we use* language, including mathematical concepts, the way we see fit; language does not force on us all the uses that could be regarded as logically possible, such as some idle and highly artificial paradoxes.

Before turning to Turing's second objection we shall briefly consider two points Wittgenstein offers in support of (iv) his insouciant lack of interest in attempts to find a consistency proof.

(1) In his conversations with Schlick and Waismann (1930), Wittgenstein made much of the idea that in mathematics we cannot meaningfully look for something if we have no systematic method of doing so. Therefore, a conjecture is said not to have a proof or disproof within an existing mathematical calculus, but can only be proven by an extension of the calculus (*PG* 359–63). By the same token, we cannot meaningfully look for, or be worried about the possibility of, a hidden contradiction (*WVC* 120). The point is repeated in the 1939 lectures: 'If there is no technique [of finding hidden contradictions], we ought not to talk of a hidden contradiction' (*LFM* 210).

This is not a very strong argument. Even if we agree to say that the proof of some innovative piece of mathematics has to go beyond the existing calculi, that doesn't make it any less real or respectable. Likewise, there could be a respectable concern about possible inconsistencies even without a method for detecting them. Moreover, it need not rule out Hilbert's idea of a methodical way of producing a proof of consistency; even though, eventually, Gödel's Second Incompleteness Theorem appears to have put paid to that ambition.

(2) Another aspect of the same idea is that an inconsistency only to be discovered in the future is not actually part of our current calculus—does not yet exist. With respect to the features of a calculus, as a normative practice, Wittgenstein adhered to the principle that *esse* is *percipi* (*RFM* 205fg).

Again, we may agree with Wittgenstein's way of putting it: what we don't know of can't be part of our mathematical practice. But that doesn't make it unreasonable to be concerned about future developments. It is not implausible to suggest that once a method for producing a contradiction has been found it will be virtually unavoidable for it to *become* part of our calculus. In his lectures Wittgenstein offers the analogy of a prison intended to be constructed in a way that no two prisoners can ever meet—:

> We could imagine that the system of corridors is very complicated—so that you might not notice that one of the prisoners can after all get by a rather complicated route into the room of another prisoner. So you have forfeited the point of this arrangement.
>
> Now suppose first that none of the prisoners ever noticed this possibility, and that none of them ever went that way. We could imagine that whenever two corridors cross at right angles, they always go straight on and never think of turning the corner. And suppose that the builder himself had never been struck by the possibility of their turning the corner at a crossing. And so the prison functions as good as gold.
>
> Then suppose someone later on finds this possibility and teaches the prisoners to turn the corner. Can we say, 'There was always something wrong with this prison?'
>
> (*LFM* 221)

We can of course say that before the discovery the prison functioned well. In that sense, there was nothing wrong with it: under the circumstances (including their ignorance of the result of turning right), prisoners were not able to meet each other. However, one can also say that, in another sense, there has been something wrong with the prison architecture all along: namely, that it has always had this weakness that once somebody discovered the possibility of turning the corner at a crossing the prison would no longer fulfil its purpose. Similarly, it would not appear implausible to worry that a calculus that works fine for the time being might become less useful through the discovery of a kind of legal move that defeats the purpose of the calculus. Another of Wittgenstein's examples is that of a game where a trick can be found by which one side is always guaranteed to win (*RFM* 203c, 373cd). Again, the mere possibility of finding such a trick doesn't make the game less enjoyable as long as the trick hasn't been found. And yet the existence of such a trick would surely be regarded as a design flaw of the game. For once it has been found and become known the game is likely to lose a lot of its appeal.

Turing's second objection puts further pressure on Wittgenstein's provocative view (iv). If an inconsistency could lead to a bridge falling down, the search for a proof of consistency would certainly appear reasonable. So how can Wittgenstein adopt such an insouciant attitude towards the possibility of an inconsistency yet to be discovered?

(1) Wittgenstein envisages only two possible sources of an engineering disaster:

> We have an idea of the sort of mistake which would lead to a bridge falling.
>
> (a) We've got hold of a wrong natural law—a wrong coefficient.
> (b) There has been a mistake in calculation—someone has multiplied wrongly.
>
> (*LFM* 211)

Chihara objects that there is a third possibility:

> [c] the logical system [the engineers] used was unsound and led them to make invalid inferences.
>
> (Chihara 1977, 334)

As already noted, this is a conflation of logic and mathematics. Engineers don't use systems of formal logic, they only use mathematics. So he should have said: 'their mathematical calculus was unsound'. One of Wittgenstein's students, Derek Prince tries to describe such a case:

> Suppose we have two ways of multiplying which lead to different results, only we don't notice it. Then we work out the weight of a load by one of these ways and the strength of a brass rod by the other. We come to the conclusion that the rod will not give away; and then we find that in fact it does give way.
>
> (*LFM* 216)

The case is similar to Wittgenstein's example of second-order exponentiation without brackets: an ambiguity that allows for two different ways of calculating leading to different results. Wittgenstein doesn't seem to deny that accidents may happen as a result of such an ambiguity; he only notes that they *need* not happen (*LFM* 217).

So, does Prince's example show that there is a third kind of cause of an engineering accident? Wittgenstein doesn't seem to think so. In a remark shortly afterwards, that may be taken to be part of his response to Prince's example (although he seems uncertain as to whether to regard Prince's case as one of a 'hidden contradiction'), he says:

> Why are people afraid of contradictions . . . inside mathematics? Turing says, "Because something may go wrong with the application." But nothing need go wrong. And if something does go wrong—if the bridge breaks down—then your mistake was of the kind of using a wrong natural law.
>
> (*LFM* 217)

That is to say, the decision to apply a certain calculus (which may allow two non-equivalent methods of multiplication—or, to take another example, a calculus that allows division by *0*) is not itself a mathematical decision. Rather, Wittgenstein suggests, it belongs to physics; just like the decision to use non-Euclidian geometry in cosmology. It amounts to the empirical claim that certain phenomena can be accurately represented or modelled by a certain conceptual system. And the choice of a conceptual system (a certain mathematical tool) for representing empirical phenomena is just an aspect of producing a mathematical formula to express a law of nature. A tool may be unsuitable for a given purpose without being intrinsically a bad tool: 'unusable for *these* purposes. (Perhaps usable for other ones.)' (*RFM* 204e).

At this point Wittgenstein continues:

> Isn't it as if I were once to divide instead of multiplying? (As can actually happen.)
>
> (*RFM* 204e)

That suggests that the choice of an unsuitable mathematical tool can also be regarded as 'a mistake in calculation'—the other possible cause of an engineering accident Wittgenstein envisages (b). That, however, would appear to be an appropriate description only where I should have known better at the time, whereas the case under discussion in the lectures is one where so far no inconsistency has been noticed; so the use of the calculus would not appear to be a mere matter of carelessness.

It seems undeniable that the choice of a suitable mathematical calculus in order to model a given range of phenomena is part of empirical science, not a priori mathematics. However, one may accept this as a feature of advanced scientific theory construction, but balk at such a claim in the case of the use of arithmetic in common-or-garden engineering. Is it not built into our arithmetic that it can be used to determine quantities of ordinary material objects or units of measurement? Given the way numbers and elementary arithmetic operations are anchored in our practice of counting material objects, it could be argued that a calculus yielding inconsistent results at that elementary level would be *mathematically* flawed.

As we saw earlier (Chapter 8.5), however, that is not Wittgenstein's view. Deviant mathematics, whose application is likely to lead to false empirical claims, is not for that matter to be regarded as itself false; it is only impractical. Wittgenstein compares inconsistent arithmetic to some of the deviant mathematical procedures already discussed, such as 'overlap' division/multiplication (*RFM* 206a), a queer way of calculating the price for wood (*LFM* 214), or elastic rulers (*RFM* 377c-f). And again, he stresses that such odd methods would 'not necessarily' get us into trouble (*LFM* 212). Yet, again, we should reply that, as a matter of empirical fact, in an environment like ours for people with anything like our form of life such alternative calculi would not be an option. We can agree with Wittgenstein that an arithmetic that yields inconsistent results is not intrinsically false, but only impractical; and yet its probable uselessness for our purposes is reason enough to regards it as a flawed piece of mathematics.

In other words, we may agree with Wittgenstein that what goes wrong when we apply an impractical calculus belongs to the side of physics (a), in the widest sense, that is, it is an *empirical* matter that we have no use for such a calculus (hence Wittgenstein's repeated insistence that things *need not* go wrong). On the other hand, since we are all agreed that we don't want our arithmetic to be inconsistent (*RFM* 377f), the subsequent discovery that what we thought was consistent does after all allow inconsistencies would make it perfectly reasonable to say that in such a case there was something wrong with our mathematics, as opposed to our scientific theories. We would indeed be inclined to regard that as a third type of possible cause of an engineering accident: an erroneous assessment not of the empirical phenomena in question, but of the calculus.

(2) So, could a bridge fall down because of a hidden contradiction? Yes, Wittgenstein does not deny the possibility (*RFM* 400b, *LFM* 217), he just thinks we would be unduly 'nervous', not to say 'silly', to be concerned about such a remote possibility (*LFM* 225).

For one thing, the paradoxes that motivated the search for consistency proofs did not occur in mathematics proper, but in the foundationalist project of reducing mathematics to logic and set theory. Varying Karl Kraus's judgement on psychoanalysis, one could say that set theory is the very source of that sceptical problem for which it is supposed to offer a solution. For Russell's paradox to lead to the collapse of a bridge we would have to imagine (as Chihara appears to have done) engineers not using ordinary mathematics, but doing their calculations in the axiomatic system of Frege's *Grundgesetze*! That indeed would be a recipe for disaster. The sheer complexity of carrying out even the most trivial sums in Frege's calculus would make errors and accidents very likely indeed—far more likely, in fact, than a derivation from a Russell type contradiction. After all, Russell's paradox has the far-fetched artificiality of the Liar construction, which is very unlikely to figure in practical jobs like building bridges.

For another thing, it should be noted that it didn't take long for the hidden inconsistency in Frege's system to be spotted—Russell wrote his famous letter to Frege in June 1902, when the second volume of *Grundgesetze* hadn't even come out yet—and it was relatively straightforward for Russell to take precautions against the paradoxes (by introducing the theory of types) in *Principia Mathematica* (*LFM* 229). Chihara objects that in fact 'it took Russell many years of intensive work' to produce his ramified theory of types (Chihara 1977, 332), but that misses the point. The main thing here was to understand how a contradiction could be derived in Frege's calculus and thus effectively to alert people to that possibility, which Russell already achieved by writing his letter. The basic idea of preventing such contradictions by restricting the formation of sets is, as Chihara has to admit, 'simple enough'. That it was still a painstaking work of many years to develop the system in full detail just reflects the complexity of the very logicist project (—in which it took Russell and Whitehead some 770 pages to derive the 'occasional useful' proposition '*1 + 1 = 2*').[3]

Thirdly, as for actual mathematics, far from getting worried and hampered by the idea that there may be hidden problems that could manifest themselves in applications, we should clearly continue to apply our mathematics as much as possible, for that is the most effective way of testing it and allowing possible problems to come to light. Feedback from applications might well provide a stimulus for refining and improving mathematical theories.

Finally, arithmetic has been in widespread successful application for millennia (cf. *RFM* 401b), and even analysis has been an enormous success in physics and engineering for centuries. If, as Turing thinks, 'it is almost certain that if there are contradictions it will go wrong somewhere', the fact that (as Wittgenstein replies) 'nothing has ever gone wrong that way yet' (*LFM* 218), makes it overwhelmingly unlikely that our arithmetic does contain any inconsistencies. Surely, if there was an inconsistency problem with arithmetic it would have surfaced over the last 2500 years or so. The idea that there may be hidden contradictions in such widely used mathematical tools—contradictions that are not just

3. It is proposition *110.643 on page 86 of Vol. II of *Principia Mathematica* (1927).

the stuff of artificially constructed philosophical puzzles (such as the Liar), but likely to cause practical problems—is exceedingly improbable: more of a kin with familiar sceptical worries in philosophy (Do other people have minds? Am I just a brain in a vat?) than with a challenge in mathematics.

It is worth noting that although Wittgenstein's views have seemed outrageous to philosophers of mathematics, engaged in some foundationalist project, such ideas are not uncommon among practising mathematicians. Thus, a similarly relaxed attitude towards hidden contradiction has been expressed by Nicolas Bourbaki, the famous collective of leading French mathematicians who since 1939 have been publishing *Éléments de Mathématique*, a systematic state of the art presentation of contemporary mathematics (11 volumes so far). In the same year in which Wittgenstein discussed his views with Turing, Bourbaki expressed their confidence that 'we will never see substantial parts of the majestic edifice of mathematics collapse due to the sudden discovery of a contradiction.'[4]

4. *En résumé, nous croyons que la mathématique est destinée à survivre, et qu'on ne verra jamais les parties essentielles de ce majestueux édifice s'écrouler du fait d'une contradiction soudain manifestée; mais nous ne prétendons pas que cette opinion repose sur autre chose que sur l'expérience. C'est peu, diront certains. Mais voilà vingt-cinq siècles que les mathématiciens ont l'habitude de corriger leurs erreurs et d'en voir leur science enrichie, non appauvrie; cela leur donne le droit d'envisager l'avenir avec sérénité* (Bourbaki 1939, E I.13)—I owe this reference to Didier Barbier.

12 Wittgenstein's Remarks on Gödel's First Incompleteness Theorem

Wittgenstein's account of mathematics as (akin to) grammar is diametrically opposed to mathematical Platonism, what he calls the idea of 'mathematics as the physics of "mathematical objects"' (MS 163, 46v). Platonist inclinations are common among mathematicians, but often philosophically naïve. However, one outstanding 20th-century mathematician with Platonist views, Kurt Gödel, did not only supplement his mathematical achievements with serious work in the philosophy of mathematics, he also regarded his most famous mathematical result—the Incompleteness Theorems (1931)—as something like a vindication of mathematical Platonism. Distancing himself from any Wittgensteinian ideas of mathematics, which he encountered between 1926 and 1928 when attending meetings of the Vienna Circle, he wrote:

> I was a conceptual and mathematical realist since about 1925. I never held the view that mathematics is syntax of language. Rather, this view, understood in any reasonable sense, can be *disproved* by my results.
>
> (Gödel, unsent letter to B.D. Grandjean; quoted in Goldstein 2005, 112)

There even appears to be biographical evidence suggesting that Gödel may have developed some of his key ideas in conscious opposition to those uncongenial Wittgensteinian views (Goldstein 2005, 73–120). So, Wittgenstein's critical remarks on the Incompleteness Theorem are of considerable interest, especially in the light of Gödel's unsympathetic attitude to Wittgenstein's views.

Gödel's famous 1931 paper 'On Formally Undecidable Propositions of *Principia Mathematica* and Related Systems I' is a contribution to the foundationalist programmes attempting to reduce arithmetic to formal logic and set theory. Gödel considers an axiomatic system *P*, similar to that of Russell and Whitehead's (1927) *Principia Mathematica*, using the vocabulary of second-order logic with identity and the successor function, and based on five suitable axioms. Then he proceeds to 'arithmetize' *P*, that is to say: he maps formulae (and sequences of formulae) onto numbers, and syntactic properties and relations onto numerical properties and relations, which can then again be expressed in *P*. Every sign, formula, and sequence of formulae is correlated with a specific 'Gödel number' in such a way that by factorisation one can work one's way backward determining which formula it is correlated with. All elementary signs of *P* are arbitrarily assigned a number: for this purpose Gödel chooses the first seven odd numbers, together with certain prime numbers > 13 for variables. Then the Gödel number of a formula is constructed as follows: the *n*th sign of the formula is correlated with the *n*th prime number (in order of increasing magnitude), each prime number is raised to a power equal to the Gödel number of the corresponding elementary sign, and finally we form the product of those numbers. For example, the tautology '$p \supset p$' would be expressed in *P* as

$$(\sim(x_1)) \vee (x_1)$$

Its elementary signs are correlated to numbers like this:

(	~	(	x_1	)	)	∨	(	x_1	)
11	5	11	17	13	13	7	11	17	13

Then the Gödel number of the formula can be calculated as follows (Hoffmann 2017, 207–9):

$$2^{11} \cdot 3^{5} \cdot 5^{11} \cdot 7^{17} \cdot 11^{13} \cdot 13^{13} \cdot 17^{7} \cdot 19^{11} \cdot 23^{17} \cdot 29^{13}$$

A proof in *P* is regarded as a sequence of formulae that is also assigned a unique Gödel number. To that purpose, we again calculate the product of the first *n* prime numbers (for a proof consisting of *n* formulae) each one raised to the power of the Gödel number of the corresponding formula of the proof. Consider, for example, the following little proof by modus ponens, with the formula codified above as conclusion:

$$(p \supset (p \vee p)) \supset (p \supset p)$$
$$p \supset (p \vee p)$$
$$p \supset p$$

If the Gödel numbers of those three formulae have been calculated to be *k*, *l*, *m* respectively, then the Gödel number of the whole proof is:

$$2^{k} \cdot 3^{l} \cdot 5^{m}$$

It is perhaps worth noting how quickly Gödel numbers become extremely big. Even for such an utterly trivial proof in elementary logic consisting of only three short formulae the Gödel number turns out to have more decimal places than there are elementary particles in the universe (Hoffmann 2017, 210).

Having introduced his system of code numbers for formulae in *P*, Gödel presents recursive definitions of various numerical predicates. In particular, he recursively defines the relation '*xBy*' that holds if and only if *x* is the Gödel number of a proof for the formula with Gödel number *y*. And finally, on the basis of '*xBy*' he can define:

$$Bew\ (x) \equiv \exists y\ (yBx)$$

In other words, a formula with Gödel number *x* is *provable* if and only if there exists a proof with Gödel number *y* of the formula with Gödel number *x*.

Finally, Gödel succeeded in constructing a formula of the form:

(G) $\sim Bew\ (n)$

whose Gödel number can be shown to be *n*. That is to say, he was able within his system *P* to construct a formula (G) that combines two characteristics:

(i) (G) has Gödel number *n*.
(ii) (G) is true if and only if the formula with Gödel number *n* is not provable in *P*.

The second characteristic can, perhaps, also be expressed more poignantly:

(ii)' (G) *says* that the formula with Gödel number *n* is not provable in *P*.

And if we take it together with (i) we can even formulate, as Gödel does in a preliminary summary of his proof, that (G) 'says about itself that it is not provable in *P*' (Gödel 1931, 175; tr.: 19)

Sticking for the time being to (i) and (ii), however, a little reflection shows that (G) must be true. For if (G) were false, then, according to (ii), the formula with Gödel number *n* would be *provable* in *P*, so it would be true. But, according to (i), the formula with Gödel number *n* *is* (G) itself. In short, if (G) were false—(G) would be true. So (G) cannot be false.

But, according to (ii), if (G) is true, the formula with Gödel number *n* is not provable in *P*—which means, according to (i), that (G) itself is not provable in *P*.

So (G) has been shown to be true, but not provable in *P*. The system *P* is incomplete.

12.1 Wittgenstein Discusses Gödel's Informal Sketch of His Proof

Wittgenstein's reactions to Gödel were written down in three sequences of remarks, dating from 1937, 1938–39, and 1941 respectively. The first group of remarks has been published as Appendix III of Part I of the *Remarks on the Foundations of Mathematics*, and has received a considerable amount of attention.[1] The 1938–39 remarks are in MS 121, pp. 76r–85v; and the 1941 remarks are from MS 163 and some of them have been published in *RFM* pp. 383–9.

Wittgenstein does not discuss Gödel's actual proof. As a *philosopher* of mathematics he does not regard it as his task to get involved in a mathematical argument (*RFM* 383gh).[2] Nor does he regard Gödel's proof as philosophically important:

> It might justly be asked what importance Gödel's proof has for our work. For a piece of mathematics cannot solve problems of the sort that trouble *us*.—The answer is that the *situation*, into which such a proof brings us, is of interest to us. 'What are we to say now?'—That is our theme.
>
> (*RFM* 388d)

And the situation into which Gödel's proof brings us, Wittgenstein appears to have conceived of according to Gödel's informal preliminary sketch of his proof, rather than considering the detailed proof itself. In particular, Wittgenstein takes up and reflects upon Gödel's idea of a sentence that 'says about itself that it is not provable' in a given axiomatic system (Gödel 1931, 175; tr.: 19). Victor Rodych regards that as 'Wittgenstein's (big) mistake', since in the actual proof the sentence in question does not, strictly speaking, say anything about itself, but is only about a certain number—which happens to be the sentence's own Gödel number (Rodych 1999a, 182). That criticism is doubly unfair. For one thing, as already noted, Wittgenstein did not intend to give a discussion or assessment of Gödel's Incompleteness Proof. In fact, in his 1937 remarks (*RFM* I, App. III)—the only one of the three texts that was at one point

1. Anderson 1958; Dummett 1959; Bernays 1959; Shanker 1988; Rodych 1999a; Floyd & Putnam 2000, 2006; Steiner 2001; Bays 2004; Priest 2004; Berto 2009a, 2009b; Lampert 2018. The most careful and detailed commentary on *RFM* I.III is an excellent paper by Wolfgang Kienzler (2008); rev. Engl. translation: Kienzler & Greve 2016.
2. Although in some cases he does consider the details of proofs (e.g. by Skolem or Cantor) in order to clarify the meaning of the results.

intended for publication—Gödel is not even mentioned by name. Rather, the issue is introduced in a quasi-fictional way:

> I imagine someone asking my advice; he says: "I have constructed a proposition (I will use '*P*' to designate it) in Russell's symbolism, and by means of certain definitions and transformations it can be so interpreted that it says: '*P* is not provable in Russell's system'. . ."
>
> (*RFM* I, 118b)

The idea is obviously taken from Gödel, but abstracting from the details of his proof and making no claim to give an exact account of it or its results. Elsewhere Wittgenstein explains that he thinks it appropriate to give a simplified version of what he finds interesting in Gödel's idea in order to bring out its philosophical importance:

> G's proof develops a difficulty that must manifest itself also in a much more elementary way.
>
> (MS 163, 39v–40r)[3]

For another thing, as already noted, it was Gödel himself who said in a summary account of his proof that a Gödel sentence *says* of itself that it is not provable in *P*. Therefore, in discussing that idea Wittgenstein can certainly not be accused of having misunderstood or misrepresented *Gödel*. That it was not Gödel's proof, but his informal discussion that Wittgenstein found philosophically interesting he explicitly says himself:

> What I'm interested in is not G's proof, but the possibilities to which G. draws our attention through his discussion.
>
> (MS 163, 37v)[4]

Even if Gödel was wrong to regard that as a reasonably accurate summary of the key element of his subsequent proof, Wittgenstein is certainly entitled to find the idea of Gödel's sketch philosophically interesting and worthy of discussion.

12.2 'A Proposition That Says About Itself That It Is Not Provable in P'

But (and this is a third response to Rodych's criticism) was Gödel wrong to summarise his proof as he did? After all, the Gödel sentence in the

3. *Der G'sche Beweis entwickelt eine Schwierigkeit, die auch in viel elementarerer Weise erscheinen muß.*
4. *Nicht der G'sche Beweis interessiert mich, sondern die Möglichkeiten auf die G. durch seine Diskussion uns aufmerksam macht.*

actual proof doesn't appear to say anything about itself; rather, it denies that a certain number has a certain numerical property. But then, the number is the Gödel number of the sentence itself and the property is defined in such a way that it applies to a number if and only if the sentence whose Gödel number it is can be proven in *P*.

To put it very succinctly, Gödel introduces an arbitrary system of numbering all formulae in a system of formalised arithmetic *P* (which we assume to be sound), in such a way that by factorising such a Gödel number one can ascertain which formula it denotes. He also defines a numerical predicate '*Bew*' such that:

(G1) '*Bew*' applies to a number *x* **iff** the sentence with Gödel number *x* is provable within *P*.

Moreover, Gödel shows how to construct a number *n* such that:

(G2) '~*Bew(n)*' is the sentence with Gödel number *n*.

He can then argue as follows:

1.	*Bew(n)* iff '~*Bew(n)*' is provable in *P*.	/From (G1) & (G2)
2.	'*p*' is provable in $P \supset p$.	/Assumption: *P* is sound
3.	*Bew(n)* ⊃ ~ *Bew(n)*.	/From (1) & (2)
4.	~*Bew(n)*.	/From (3)
5.	'~*Bew(n)*' is not provable in *P*.	/From (1) & (4)

In a word, Gödel has presented a sentence:

(G) ~*Bew(n)*

—that is both true (conclusion: 4) and not provable in *P* (conclusion: 5). Call this sentence (G) a *Gödel sentence*, call '*Bew*' a *Gödel predicate*, and call the system of stipulations entailing (G1) and (G2) *Gödelisation.*

(a) Gödel predicates are defined in such a way that their application to a Gödel number is *equivalent* to a claim about the formula correlated to that Gödel number.
(b) Gödelisation makes it possible for a sentence to say something about *its own* Gödel number.

But if Gödelisation (the system of Gödel numbering and the definition of Gödel predicates) allows us to *infer* such a self-referential claim, that claim can plausibly be said to be *implicit in the meaning* of (G), the Gödel sentence.

So it can, after all, be said (as Gödel himself does in his informal proof sketch) that a Gödel sentence *says* of itself that it is unprovable;[5] and is thus at least comparable to the Liar Paradox. Hence, when Wittgenstein discusses Gödel's result in terms of such self-reference (*RFM* 118b–123; MS 121, 78v–85v) he is not only justified in as much as he follows Gödel's own summary[6]—and so is certainly discussing an idea put forward by Gödel—, but also with respect to Gödel's detailed proof it is quite appropriate to say of a Gödel sentence that 'by means of certain definitions and transformations it can be so interpreted that it says' of itself that it is not provable in *P* (*RFM* 118b).[7]

12.3 The Difference Between the Gödel Sentence and the Liar Paradox

However, there is an important difference between the case of outright self-reference and Gödel's set-up of self-reference via Gödel numbers. Consider:

(GL) This sentence cannot be proven.

If it could be proven it would be false (as it claims to be unprovable), but also true (as proven). So a proof would lead to contradiction. So it cannot be proven. So it must be true.

But then its truth is hollow: it fails to express any actual proposition. It is a case of reference failure, which arguably is also the ultimate defect of the Liar sentence (cf. *Z* §691; Ryle 1950).

By contrast, the Gödel sentence has, as it were, two layers of content. It does not *only* claim its own unprovability. Rather, first of all it says that a certain number does not have a certain numerical property (or does not belong to a certain set of numbers); which only at a second step turns out to mean that the sentence itself is unprovable.

5. Gödel writes: 'We therefore have before us a proposition that says about itself that it is not provable in *P* . . . initially it [only] asserts that a certain well-defined formula . . . is unprovable. Only subsequently (and so to speak by chance [*zufällig*]) does it turn out that this formula is precisely the one by which the proposition itself was expressed' (Gödel 1931, 175; tr.: 19, 42). The parenthesis is of course somewhat inaccurate, for logical implications even if drawn only 'subsequently' are not for that matter merely accidental or contingent.
6. Wittgenstein was well aware that those remarks were not the actual proof, calling it 'Gödel's introductory casual line of proof' [*Gödels einleitende beiläufige Beweisführung*] (MS 126, 127).
7. For a detailed argument to show that Gödel's actual proof does indeed depend on such meta-mathematical interpretations, see Lampert 2018, 338–43.

This, it seems to me, is Gödel's most plausible defence against the objection that (G) is a version of the paradoxical Liar sentence and hence equally faulty. It is not that the Liar's self-referentiality has to 'depend on contextual, empirical factors' (Berto 2009b, 9), for defining and using a suitable reflexive pronoun makes it straightforwardly analytic. Nor is it, as Gödel suggests in a footnote, that the self-referentality of (G) is only contingent (*zufällig*). There are no contingencies in logic or maths.—No, the difference is that unlike the Liar sentence, (G) cannot be dismissed as a case of reference failure: as failing to express a proposition.

12.4 Truth and Provability

Wittgenstein was not interested in Gödel's proof, but only in 'the *situation*, into which such a proof brings us' (*RFM* 388d). That might be taken to suggest that he was interested in the proof's result, the First Incompleteness Theorem; but, as a matter of fact, he wasn't—:

> The mathematical fact that here's an arithmetical proposition which in P can neither be proven nor demonstrated to be false doesn't interest me.
>
> (MS 163, 37v–38r)[8]

Wittgenstein accepts the incompleteness of a formalised arithmetic as a mathematical fact, but takes no philosophical interest in it. He is interested in Gödel's method, but not in his results: '*What* he proves isn't what concerns us' (MS 163, 40v).[9]

This is not as surprising as it might appear. Remember, to begin with, that Wittgenstein never accepted the fundamental assumption on which Gödel's considerations are based, namely that arithmetic can and ought to be reconstructed in terms of formal logic and set theory, an axiomatic system *P*, along the lines of *Principia Mathematica*. As laid out in Chapter 3, Wittgenstein rejected Frege's and Russell's logicism for a number of reasons. But if arithmetic cannot be identified with its logical or set theoretical reconstruction in an axiomatic system such as *P*, the incompleteness of the latter must not be confused with the incompleteness of arithmetic.

Gödel's Incompleteness Theorem is sometimes said to have established that mathematical truth does not coincide with provability (cf. Connes 2000). But in fact, all the theorem could be said to show is that mathematical truth does not coincide with provability *in P*, or in any one

8. *Die mathematische Tatsache, daß hier ein arithmetischer Satz ist, der sich in P nicht beweisen noch als falsch erweisen läßt, interessiert mich nicht.*
9. *Was er beweist, geht uns nichts an, ...*

axiomatic system. Unlike Hilbert, Wittgenstein never thought it did. On the contrary, he emphasises that mathematics is a motley of various techniques of proof (*RFM* 176c), resisting reduction to a single calculus.

Gödel himself did not claim that his famous proof had established the existence of unprovable mathematical truths. Rather in agreement with Wittgenstein on this point, he merely took himself to have shown up the limitations of a mechanical formalism in mathematics:

> My theorems only show that the *mechanization* of mathematics, i.e. the elimination of the *mind* and of the *abstract* entities, is impossible. . . . I have not proved that there are mathematical questions undecidable for the human mind, but only that there is no *machine* (or *blind formalism*) that can decide all number theoretical questions (even of a very certain special kind).
>
> (Gödel 2003, 176)

However, Gödel did think that mathematical truth could not coincide with provability, but for different reasons, though closely related, which he explained in a letter:

> it follows from the correct solution of the semantic paradoxes i.e., the fact that the concept of 'truth' of the propositions of a language *cannot* be expressed in the same language, while provability (being an arithmetical relation) *can*. Hence true ≠ provable.
>
> (Gödel, letter to Balas (around 1970), Vol. IV, p. 10; in Mancosu 2004, 246)

Gödel's idea seems to be that semantic paradoxes, such as the Liar, can only be avoided by a careful distinction between object language and meta-language, along the lines Tarski suggested for artificial formal languages (Tarski 1937; cf. Sainsbury 1995, 118–21). If truth claims can only be made about sentences in a distinct, lower-level language, the self-referentiality required for the Liar Paradox is made impossible. However, *avoiding* a paradox by banning the use of language that allows it to be formulated is not the same as *solving* it (—just as a chess problem cannot be solved by changing the rules of chess on which it depends). It only means brushing it under the carpet. The fact remains (and Tarski was well aware of it (1936, 406)) that natural languages are not stratified into object and meta-language, and it is perfectly licit and common to make truth claims about statements in the same language. What is more, such predications of truth are an extremely useful and well-functioning device of our natural languages, which we have no reason to abolish just because in an idle moment some philosophers managed to use that predicate to construct a mildly entertaining, but utterly pointless puzzle (*RFM* 120a). In any case, it is not the application of the truth predicate to statements of the same language that leads to paradox, but only the very peculiar uses where it is applied to the very statement in which it occurs (which therefore can be argued to fail to express any proposition).

Gödel is wrong to assert that truth cannot be expressed in the same language. It can in English; and even in a calculus one can easily introduce a truth operator:

p	Tp
T	T
F	F

Moreover, it is inaccurate and misleading to say that provability is an arithmetical relation. For one thing, the relation that Gödel manages to define in his system P is merely provability-in-P. He can plausibly speak of mathematical truths not provable in P only because they can be proven by different means. For another thing, the definition of provability-in-P implies the notion of truth in P, for 'to prove' means: to show to be true. That is to say, unlike the purely syntactical concept of derivability, provability is a semantic, or epistemic, concept. Deriving both p and $\sim p$ in an inconsistent calculus is not a proof of p and $\sim p$. We accept a derivation as proof only if we regard the calculus as sound. Thus the concept of provability remains linked to that of truth and cannot be defined independently.

In fine, all Gödel is entitled to conclude is: 'mathematically true $\neq$ provable-in-P'. Nothing he shows prevents us from insisting that (cf. *RFM* 118c):

(i) true in P = provable in P;
(ii) mathematically true = mathematically provable (or accepted as an axiom).[10]

Against (i), it might be objected that Gödel's sentence (G) is an instance of a formula of P that is true, but not provable in P, since its truth is established by meta-mathematical reasoning (viz.: 'if it were false, it would be false that it's unprovable; so it would be provable; so it would be true; so it cannot be false, but must be true'). However, Wittgenstein points out that a truth that can be written down with the vocabulary of a given axiomatic system need not for that matter be true in that system (*RFM* 118–9). It may, for example, be possible to write down an empirical truth in the symbolism of *Principia Mathematica*; such an empirical truth would obviously not be a truth *of Principia Mathematica*, i.e. a logical truth.

10. Note that not every possible calculus is acceptable as mathematics; not every rule-governed derivation of signs from other signs is a mathematical proof. A mathematical proof, according to Wittgenstein, must develop concepts that have a use outside mathematics, or at least link up with such concepts (*RFM* 257de).

Or again, imagine a calculus *Q* covering only a fragment of propositional logic: with the rule of conjunction introduction, but nothing like conjunction elimination. Then it may be possible in *Q* to derive as theorems certain conjunctions, but not their individual conjuncts. Invoking the natural interpretation of conjunction in *Q*, we may then argue that if 'φ . ψ' is true, so surely are both 'φ' and 'ψ'. But again, such argument takes us beyond the system. In *Q*, 'φ . ψ' may be a theorem, while the individual conjuncts are not.

In fine, as Gödel does not succeed in disproving the equivalence between truth and provability in mathematics, he cannot be said to have established mathematical Platonism.[11]

12.5 Gödel's Kind of Proof

If Wittgenstein was not much interested in Gödel's results, in the incompleteness of a calculus such as *P*, what *was* his interest in Gödel's article? What did he mean by 'the *situation*, into which such a proof brings us' (*RFM* 388d)?

What is most striking about Wittgenstein's remarks on Gödel is what might appear as a certain stubborn perversity in repeatedly asking the very question that, one should think, Gödel's proof has ruled out as impossible. Gödel has established that a certain sentence (G) cannot be proved in *P* (assuming that *P* is consistent); yet Wittgenstein keeps asking: 'What if we *did* find a proof for (G) in *P*?'.

> Now how am I to take [G] as having been proved?
>
> (*RFM* 121e)[12]

> If we had then derived *the arithmetical proposition* [G] from the axioms according to our rules of inference, then *by this means* we should have demonstrated its derivability, but we should also have proved a proposition which, by that translation rule, can be expressed: this arithmetical proposition (namely ours) is not derivable.
>
> (*RFM* 387b)

11. Cf. Hacking 2014, 202: 'as an argument for the existence of an archaic arithmetical reality, with all its truths intact, [invoking incompleteness] begs the question. If you don't think of that reality as already given, then for any consistent and adequate axiom system, Gödel exhibits one sentence which we have grounds to call true, but not provable in that system. This does not show that there is an actual totality of all arithmetical truths'. The fact that for any such system there is something than cannot be proved in that system doesn't show that there is something that cannot be proved in any system.
12. Wittgenstein speaks of a sentence 'P' not provable in Russell's system. Since the letter 'P' was also Gödel's name for the axiomatic system in question, I've changed Wittgenstein's sentence letter to 'G', as the conventional label of the Gödel sentence.

> But now: 'I am not provable in this way k'. Suppose we could derive this sentence in that way . . . .
>
> (MS 163, 32r)[13]

> If by basic logical and arithmetical principles you had derived a mathematical sentence whose most natural application appeared to be making the derivation of the derived sentence appear hopeless . . . .
>
> (MS 163, 32v–33r)[14]

Why does Wittgenstein focus so much on considering the possibility of what (G) tells us is impossible? Two remarks shed some further light on Wittgenstein's peculiar approach:

> *What* [Gödel] proves isn't what concerns us, but we have to consider that *kind of* mathematical *proof*.
>
> (MS 163, 40v)[15]

> The unphilosophical nature of Gödel's article shows itself in the fact that he doesn't see the relation between mathematics and its application. Here he has the slimy concepts of most mathematicians.
>
> (MS 124, 115)[16]

Wittgenstein finds a new *method* of mathematical proof in Gödel's article, which is of interest to a philosopher of mathematics. Moreover, he regards the issue of *applicability* as relevant to Gödel's proof in a way that Gödel himself doesn't seem to have realised.

The peculiarity of Gödel's kind of proof is that the resulting theorem (viz., that a certain formula of *P* is not decidable in *P*) is not actually derived in *P*. Rather, Gödel constructs a formula (G) in such a way that, roughly speaking, it can be interpreted to say of itself that it is not provable in *P*. Given that self-referential interpretation of (G), it can then be argued that (G)—although not derived in *P*—must be true. In short, in his proof Gödel uses not a step-by-step derivation according to the rules

13. *Aber nun: 'Ich bin nicht auf die Weise k beweisbar.' Nehmen wir an wir können den Satz auf diese Weise ableiten; . . .*
14. *Hättest Du einen mathematischen Satz aus logischen und arithmetischen* [33r] *Grundprinzipien abgeleitet, dessen natürlichste Anwendung zu sein schiene das Ableiten des abgeleiteten Satzes als hoffnungslos darzustellen, . . .*
15. *Was er beweist, geht uns nichts an, aber wir müssen uns mit dieser mathematischen Beweisart auseinandersetzen.*
16. *Das Unphilosophische an Gödels Aufsatz liegt darin, daß er das Verhältnis der Mathematik und ihrer Anwendung nicht sieht. Er hat hier die schleimigen Begriffe der meisten Mathematiker.*

of the calculus, but the *interpretation of an unproven formula* in order to show that that formula must be true (cf. Lampert 2018).

Remember that according to Wittgenstein mathematics is a system of quasi-grammatical norms that have to be endorsed by proof. Mathematical truth is the normative force given to a formula as the result of a proof. An unproven formula may serve as an inspiration to look for a proof, but in the meantime it has no normative force. If, as in Gödel's case, an interpretation of an unproven formula allows us to argue that it must be true—we have thereby provided a proof for that formula, albeit a different kind of proof from the ones inside *P*.

But now what is it we have proved by proving (G)? We are supposed to have proved that (G) cannot be proved within *P*. And now we come to what Wittgenstein accuses Gödel of having neglected: the *application* of (G). According to Wittgenstein, the application of an impossibility proof is a psychological one: it is to dissuade people from trying to do something.

> If we really prove that the heptagon cannot be constructed, it should be a proof which makes us give up trying—which is an empirical affair. And similarly with proving that a certain proposition is not provable.
>
> (*LFM* 56)

A significant point here is that, in such a case, Wittgenstein always considers the possibility that we are confronted with something that *appears* to be a proof of the proposition in question, or a construction of the figure in question. For example, it is easy enough to imagine someone coming up with a complicated drawing resulting in a figure with seven sides that, according to our measurements, are all equal. And similarly, we could imagine a very long derivation in *P* leading to the formula '*~Bew(n)*'.—One may be inclined to object that according to Gödel's proof that is logically impossible and so cannot even be imagined, but Wittgenstein replies that what we need to imagine is only that we *appear* to encounter a proof of (G), which may in fact be based on a hidden miscalculation that somehow we haven't yet been able to identify (*RFM* 388b).

Now, Wittgenstein's criterion for a successful impossibility proof is that in such a situation (of encountering an apparent counter instance) the impossibility proof gives us a 'forcible reason' to reject the alleged counter proof (*LFM* 56; *RFM* 120c). In one sense that is trivial. Of course, a proof that *p* gives you a reason to believe that *p*; hence a proof that something is mathematically impossible gives you a reason to reject any claim that it has been achieved. But what Wittgenstein has in mind goes further than that. What he means is that proving the impossibility of achieving *X* must give you 'a very much clearer idea' of what would be involved in achieving *X* (*LFM* 87). For example, the proof that it is impossible to construct a heptagon with ruler and compass enhances our understanding of the method of constructing an *n*-sided polygon with ruler and compass in such a way that we can see that it cannot be done for $n = 7$. Here we encounter once again Wittgenstein's idea that a mathematical proof amounts to a change of a concept. 'The problem [of constructing a heptagon] arose because our idea at first was a different idea of constructing the n-gon, and then was *changed* by the proof' (*LFM* 89). However, as suggested earlier (in Chapter 10.1), it is more plausible to regard this as an increase in our understanding of the implications of a given concept (and its relations to other concepts), rather than a change of concept. Yet it is true that such a better understanding (or 'idea') of a concept may lead to a new technique, which makes it easy to distinguish the cases where the concept is applicable from those where it isn't.[17]

Thus, for Gödel's proof to give us a good reason to regard the formula '*~Bew(n)*' as undecidable in *P*, the proof should enrich our understanding of the techniques of derivation in *P*—but it does not. Wittgenstein's complaint could perhaps be put by saying that Gödel's proof yields a result, but no understanding.

12.6 Wittgenstein's First Objection: A Useless Paradox

Wittgenstein's complaint that Gödel's proof fails to provide a 'forcible reason' to accept that (G) could not be derived in *P* is developed further along two distinct lines. The first one emphasises the paradoxical anomaly of (G). Wittgenstein writes:

> 14. A proof of unprovability is as it were a geometrical proof; a proof concerning the geometry of proofs. Quite analogous e.g. to a proof that such-and-such a construction is impossible with ruler and compass. Now such a proof contains an element of prediction, a physical element. For in consequence of such a proof we say to a man: "Don't exert yourself to find a construction (of the trisection of an angle, say)—it can be proved that it can't be done". That is to say: it is essential that the proof of unprovability should be capable of being applied in this way. It must—we might say—be a *forcible reason* for giving up the search for a proof (i.e. for a construction of such-and-such a kind).
>
> A contradiction is unusable as such a prediction.
>
> [*RFM* 120cd]

17. Thus, for example, Gauß showed in his *Disquisitiones arithmeticae* (1801) which of the infinitely many possible regular polygons can be constructed and which cannot (Boyer & Merzbach 1989, 563).

The second paragraph's laconic verdict can appear somewhat puzzling. Consider another impossibility proof, for example, Euclid's proof that there is no greatest prime number. That proof proceeds by assuming the existence of a greatest prime number and then deriving a contradiction—which seems perfectly suitable for convincing us to give up the search for a greatest prime number. So, as Paul Bernays remarked, it is rather strange that Wittgenstein should say that a contradiction is unusable for such a purpose: 'proofs of impossibility in fact always proceed by the deduction of a contradiction' (Bernays 1959, 523).

Wolfgang Kienzler, in his perceptive commentary on *RFM* I, App. III, offers two replies to Bernays's objection (Kienzler 2008, 179–80). First, in Gödel's proof there is no specific mathematical proposition from which we are dissuaded. Secondly, the contradiction in question is nothing but a 'useless paradox'. The latter seems to me the right reply (to which I shall return presently), whereas the former point is less convincing. Kienzler writes that in Gödel's reasoning there is '*not* some (mathematical) *thing* which was then shown to be unprovable—but there is, rather, nothing (except the contradiction itself) . . . . Similarly, it is not the case that people had previously tried to prove Gödel's formula but then Gödel showed that it could not be done' (Kienzler & Greve 2016, 104).

However, it is not clear to me why that should amount to an objection to Gödel. If it is obvious that (G) is unprovable, so that no-one would even try to prove it,—so much the better. But what was not obvious before, is that there are undecidable sentences in *P*; and so people may have tried to find proofs for the completeness of *P*. Presenting an obviously unprovable sentence of *P* is just what it takes to dissuade people from looking for completeness proofs. Moreover, Wittgenstein does not in fact seem to think that (G) is *obviously* unprovable, for he keeps considering the possibility of finding a proof of (G). And that, really, there is no 'mathematical thing' there, is just as it should be: likewise, there is no such thing as a greatest prime number, which is what Euclid's proof by contradiction shows. That is the situation with all impossibility proofs: they show up a conjecture to be inconsistent.

Kienzler's second point, on the other hand, seems to me spot on. As the context makes clear, what Wittgenstein has in mind when in *RFM* 120d he speaks of a 'contradiction' is not the logicians' contradiction '*p* . ~*p*' that figures in a *reductio ad absurdum* proof. Rather, it is the kind of contradiction involved in the Liar Paradox (*RFM* 120a): the contradiction implicit in a semantic paradox ('if provable then false'). This, then, is Wittgenstein's first objection to Gödel. Like the Liar sentence, (G) with its self-referential interpretation is just a curious anomaly: mathematical language gone on holiday (cf. *PI* §38). Because of its paradoxical artificiality it remains mathematically barren: it cannot really convince us of anything. Thus it could be suggested that in a more practical sense the question of the completeness of *P* remains open: perhaps all proper sentences—those with some useful mathematical content—are decidable in *P*. Unlike the '*p* . ~*p*' in a *reductio* argument, (G) is an *idle* contradiction to which we need not pay any attention.

Gödel himself was not impressed by this objection. In a letter to Karl Menger he responded indignantly:

> As far as my theorem about undecidable propositions is concerned, it is indeed clear from the passages you cite that Wittgenstein did *not* understand it (or pretended not to understand it). He interprets it as a kind of logical paradox, while in fact it is just the opposite, namely a mathematical theorem within an absolutely uncontroversial part of mathematics (finitary number theory or combinatorics). Incidentally, the whole passage you cite seems nonsense to me.
>
> [Gödel, letter to Karl Menger, 20th May 1972; quoted in Ramharter 2008, 69–70]

Of course Gödel is right that his proof is not at all like the Liar Paradox *resulting* in a contradiction. However, the 'contradiction' or paradox Wittgenstein demurs at in *RFM* 120d is more plausibly located in the semantic set-up of Gödel's proof, as described in an earlier remark:

> 9. For what does it mean to say that *[G]* and "*[G]* is unprovable" are the same proposition? It means that these *two* English sentences have a *single* expression in such-and-such a notation.
>
> [*RFM* 119b; proposition labels adapted]

In other words, Wittgenstein takes exception to the way Gödel's proof is based on a paradoxical notational ambiguity. (G) and '(G) is unprovable' are not formally contradictory, but engender a quasi-contradictory tension once we consider their epistemic support: what proves (G) disproves '(G) is unprovable'; and what proves '(G) is unprovable' discredits (G). Hence, representing both by one and the same expression engenders paradox. It results in a sentence such that proving it would be showing it to be false—making a nonsense of the concept of proof.

12.7 Wittgenstein's Second Objection: A Proof Based on Indeterminate Meaning

Another, more specific, objection to Gödel is developed in the following three remarks of *RFM* I, App. III (§§15–17). First, Wittgenstein reminds us of his view (discussed in Chapter 10.2 above) that the meaning of a mathematical proposition is (at least partly) determined by its proof:

> The proof is part of the system of operations, of the game, in which the proposition is used, and shews us its 'sense'.
>
> [*RFM* 121a]

> Now, in order to see *what* has been proved, look at the proof.
> [*RFM* 121e]

It follows that while unproven, (G) has no clear mathematical meaning. In particular, it is not good enough to explain that (G) carries the meaning that '(G) is unprovable', for it remains to be determined what exactly that means: what is to count as a proof (*RFM* 121d). But then, if (G)'s meaning remains somewhat indeterminate as long as it has not been given the status of an acknowledged mathematical proposition—if 'its sense is still veiled' (*RFM* 121d) —, we cannot take (G)'s meaning as a basis of a meta-mathematical proof of (G). In other words, from Wittgenstein's point of view, there is a vicious circularity in Gödel's meta-mathematical proof of the truth of (G): it must presuppose the meaning of (G) in order to prove—and hence to determine the meaning of—(G).

Because of the initial indeterminacy of (G) we can still imagine finding a proof of (G); and only then, depending on what that proof looks like, can we give a precise meaning to (G). In the following remark (§17), Wittgenstein considers some possibilities.[18] The upshot may be that in the light of finding what strikes us as a proof of (G) in *P* after all, we decide to retract our initial interpretation of (G) as implying its own unprovability (*RFM* 121e; cf. Lampert 2018, 326).

18. For a detailed elucidation of the possibilities considered, see Kienzler 2008, 182–4.

13 Concluding Remarks

Wittgenstein and Platonism

Naturally the biggest question in the philosophy of mathematics is: *What is mathematics?* That is to say: *what is the status and meaning of mathematical propositions?* Probably the most popular answer has always been the one given by Platonists: that mathematical propositions are descriptions of independently existing abstract objects. Wittgenstein's philosophy of mathematics is radically anti-Platonist. Already in the *Tractatus* Wittgenstein offered an account of mathematics as non-descriptive, so a fortiori not involving descriptions of Platonic mathematical objects. He was well aware of the Platonic implications of set theory, based on Cantor's view of infinite totalities, and therefore sketched an alternative logicist account of arithmetic, not in terms of classes or sets, but in terms of operations (Marion 1998, 2–3). And it remains a leitmotif of Wittgenstein's philosophy of mathematics all through his career that mathematics is algorithmic and not descriptive (*WVC* 106, *PR* 188); that the 'mathematician is an inventor, not a discoverer' (*RFM* I §168: 99). In response to the what-is question Wittgenstein offers an account of mathematics as (something like) grammar that invites us to look at mathematics in a way that eschews even the temptation to resort to any form of Platonism.

In fact, Wittgenstein's anti-Platonism is not just a particular view of mathematics; its roots lie in his most fundamental insight in the philosophy of language, namely: that linguistic meaning is determined by use (*PI* §43) and that different linguistic expressions often are used in markedly different ways disguised by superficial syntactic similarities (*PI* §§8–17). Hence we constantly need to resist the natural descriptivist prejudice (illustrated by the quotation from Saint Augustine at the beginning of *Philosophical Investigations*), according to which all language is a matter of naming and describing objects. Cartesian Dualism is a consequence of such a naïve belief in linguistic uniformity (critically discussed in *PI* §§243–315, the so-called private language argument). If all nouns are construed as names of objects, then nouns such as 'pain', 'feeling', 'thought', or 'wish', must presumably be names of *inner* objects: occurrences in the privacy of one's consciousness.[1] In a similar way, mathematical Platonism results from construing number words as names of objects—abstract objects existing outside space and time. In both cases a grammatical difference (a difference in the use of certain words) is misunderstood as an ontological difference (a difference in the objects denoted), due to a naïve insistence on a simplistic denotational picture of the function of language.

Already in the first section of the *Investigations* Wittgenstein points out how, at the most elementary level, mathematical Platonism arises as the result of semantic prejudice, a referentialist (or denotational) account of meaning. He imagines a simplified language game to illustrate how we operate with words (cf. *PI* §449); how we use sentences as instruments to achieve various purposes (*PI* §421), for example: to buy some apples. The upshot is that an imagined Platonist referent would be irrelevant to the actual use of a number word (*PI* § 1; quoted in Chapter 3.3). Admittedly, in *PI* §1 the point is made only with respect to the most elementary arithmetical device, counting, but there is no reason why it should not hold for more advanced mathematics as well. That is at any rate the view of the Cambridge mathematician and Fields medallist Timothy Gowers who describes Platonism, using one of Wittgenstein's metaphors, as an idle wheel, not part of the mechanism (cf. *PI* §271), remarking that the work of a mathematician holding Platonist beliefs would not on that account differ in any way from that of an anti-Platonist colleague or a mathematician that regarded such metaphysical questions as altogether pointless (Gowers 2006, 198).[2]

In his seminal paper 'Mathematical Truth' (1973), Paul Benacerraf gives an account of the philosophical struggle with Platonism in mathematics that, although Benacerraf remains sympathetic to Platonism, can be seen as a vindication of Wittgenstein's anti-Platonist view. Benacerraf explains very plausibly how Platonism appears to be forced upon us by what he calls the 'standard view' of semantics, namely that language, including mathematical language, denotes and describes objects, and he explains further how the ensuing view of abstract mathematical objects gets into conflict with our standard epistemology. For how can we have knowledge about objects that are not accessible to sense perception?

1. For a detailed account of Wittgenstein's attack on this 'inner object' model of the mind, see Schroeder 2006, 181–233.
2. Wittgenstein might have objected, though, that Platonist mathematicians were more likely to waste their time with set theory (cf. *RFM* 264d), but perhaps it could be argued that his notorious dislike of set theory was directed not at the calculus as such, but as its misguided application (*RFM* 260a) (to give a foundation to arithmetic) and its common interpretation (as a description of the actual infinite (*PG* 368)). Cf. Marion 1998, 15–19.

The most natural Platonist response was given by Gödel (1947, 271): beside sense perception we must acknowledge another kind of perception, namely 'mathematical intuition'. Yet as Benacerraf rightly remarks, that is just putting a label on a way of acquiring knowledge that we do not really understand (1973, 497). According to Benacerraf, no philosophy of mathematics has so far succeeded in satisfying both the semantic requirement (of construing language as denoting and describing objects) and the epistemological requirement (of explaining our knowledge of the objects in question), but he appears to be optimistic that eventually such a philosophical theory could be found. From Wittgenstein's point of view, however, Benacerraf's dilemma is just a neat illustration of how the pervasive prejudice that all language must fit the same referentialist pattern (the naïve Augustinian picture of language) gets us into trouble, engendering an irresolvable dilemma. Reject the naïve belief in referentialism and the homogeneity of language—and Benacerraf's problem doesn't arise. What is remarkable about Benacerraf's paper is that he fully understands that it is only the insistence on a homogenous referentialist semantic theory that produces the problem; and yet he can't free himself from that insistence, so overwhelmingly natural does he find the idea that all language denotes and describes objects and so implausible does he find any alternative account of mathematics.

We saw in the previous chapter that, contrary to a common belief, Gödel's First Incompleteness Theorem does not provide any new argument in favour of Platonism. More specifically, since Gödel's result concerns only provability in a given axiomatic system: it does not show that truth cannot be identified with provability tout court. However, Gowers presents a different objection to identifying mathematical truth and provability, which might be thought to speak in favour of Platonism.

> consider the statement 'somewhere in the decimal expansion of π there is a string of a million sevens'. Surely, one feels, there is a fact of the matter as to whether that is true or false, even if it may never be known which.
>
> What is it that makes me want to say that the long string of sevens is definitely either there or not there—other than a general and question-begging belief in the law of the excluded middle? Well, actually I am tempted to go further and say that I believe that the long string of sevens *is* there, and I have a definite reason for that stronger belief, which is the following. All the evidence is that there is nothing very systematic about the sequence of digits of π. Indeed, they seem to behave much as they would if you just chose a sequence of random numbers between 0 to 9. This hunch sounds vague, but it can be made precise as follows: there are various tests that statisticians perform on sequences to see whether they are likely to have been

> generated randomly, and it looks very much as though the sequence of digits of π would pass these tests. Certainly the first few million do. One obvious test is to see whether any given short sequence of digits, such as 137, occurs with about the right frequency in the long term. In the case of the string 137 one would expect it to crop up about one thousandth of the time in the decimal expansion of π. If after examining several million digits we found that it had in fact occurred a hundredth of the time, or not at all, then we would be surprised and wonder whether there was an explanation.
>
> But experience strongly suggests that short sequences in the decimal expansions of the irrational numbers that crop up in nature, such as π, e or the square root of 2, *do* occur with the correct frequencies. And if that is so, then we would expect a million sevens to occur in the decimal expansion of π about $10^{-1000000}$ of the time—and it is of course no surprise that we will not actually be able to check that directly. And yet, the argument that it does eventually occur, while not a proof, is pretty convincing. . . .
>
> This . . . raises difficulties for those who are too ready to identify truth and provability. If you look at actual mathematical practice, and in particular at how mathematical beliefs are formed, you find that mathematicians have opinions long before they have formal proofs. When I say that I think π almost certainly has a million sevens somewhere in its decimal expansion, I am not saying that I think there is almost certainly a (feasibly short) *proof* of this assertion—perhaps there is and perhaps there isn't. So it begins to look as though I am committed to some sort of Platonism—that there is a fact of the matter one way or the other and that that is why it makes sense to speculate about which.
>
> (Gowers 2006, §6; cf. Davis & Hersh 1982, 363–9)

A Wittgensteinian response to this consideration is twofold, according to the two key ideas in Wittgenstein's thinking, which I called the 'grammar view' and the 'calculus view'.

First, the grammar view, regarding mathematics as a system of norms of representation, is the most distinctively Wittgensteinian contribution to the philosophy of mathematics and his primary line of rebutting Platonist inclinations. Mathematics does not describe a world of mathematical objects since, like a system of grammatical rules, it doesn't describe anything, but regulates inferences among our empirical propositions. However, the grammar view does not require all such quasi-grammatical norms to be established by proof. We could imagine them to be based on the authority of priests or a caste of wise men. More realistically, such norms can be (and indeed often were) abstracted from trial and error, as Wittgenstein explicitly envisaged speaking of empirical observations 'hardened into rules' (*RFM* 324b). From this point of view, Gowers's description of how mathematicians can acquire mathematical convictions without any formal proof is not an embarrassment. All that matters is that the mathematical community can be prevailed upon to accord some normative status to certain mathematical propositions, be that on the basis of formal proof, reasonable empirical conjecture, or even religious speculation.

Secondly, as we saw, Wittgenstein also defends the view that mathematical results must be established by calculation. It is this 'constructivist' line in Wittgenstein's anti-Platonism that would appear challenged by Gowers's observations. However, Gower himself proceeds to suggest a plausible reply. Even though it is essential to our mathematical practice that new mathematical claims are established by a demonstration of correctness, we may have to note that what mathematicians accept as 'proof', that is, as reasonable grounds on which to regard a mathematical proposition as true, need not be a formal derivation in an axiomatic system. Gödel gave an example of how something can be shown to be correct without receiving a formal proof, and Gowers illustrates how mathematicians can be convinced, at least tentatively,[3] by statistical or probabilistic considerations, what he calls a 'reasonable-sounding heuristic argument'.

In fact, such a realistically flexible approach to what we regard as a convincing argument in mathematics fits very well with Wittgenstein's anthropological view of mathematics. If mathematical proofs are to be understood as arguments that convince us to accept something as a norm of representation (cf. *RFM* I §33: 50; §63: 61), it is only to be expected that there may be different degrees of conviction, as there are in other forms of rational discourse. Hence, there are mathematical conjectures that mathematicians have reasons to accept as true—even where they don't expect a formal proof to be forthcoming—and that scientists apply successfully, which should therefore be regarded as belonging to our corpus of mathematics, even though their status remains more precarious than that of rigorously provable mathematics.

It is Wittgenstein's emphasis of the anthropological dimension of mathematics as essentially a human practice that sets him apart from all major schools in the philosophy of mathematics.

Platonism sees mathematics not from a human, but from God's perspective. Mathematical facts are thought to exist regardless of human understanding; they exist, as it were, in the mind of God (*PI* §352). Moreover, in order to maintain the rationalist idea that mathematics provides

3. Gowers gives another example of a conjecture that seems 'certainly true': the twin-prime conjecture, noting that he is less confident about this one than about the statistically normal distribution of number sequences in π (2006, §6).

a description—indeed, the most indubitable description—of reality (what Davis and Hersh have called 'the Euclid myth' [1982, 322–30]), Platonism has to postulate a distinct realm of reality, like the transcendence of religious metaphysics, a kingdom of eternal truth. Wittgenstein's objection is that what is transcendent cannot explain our human practice.

Formalism rightly emphasises the human practice of producing mathematical proofs and is in many ways a plausible alternative to Platonism. Wittgenstein's objections to Hilbert's formalism and the concern with consistency proofs were discussed in Chapter 11. Where Wittgenstein differs also from later versions of formalism is in his insistence that application is essential to mathematics too (cf. Klenk 1976, 26–8). This is a crucial aspect of the idea of mathematics as grammar: that it regulates language applied outside pure mathematics. Not that every single theorem must be applied, or even applicable, but what distinguishes mathematics on the whole from a mere network of games is that its elementary parts are firmly established in empirical applications and its more advanced parts at least tend to be geared towards, or considered for, possible applications. As explained above, it is this second key idea in Wittgenstein's account that allows him the flexibility to accommodate aspects of mathematical practice that embarrass the strict formalist identification of mathematical truth with provability, such as the cases of tentative mathematical reasoning presented by Gower or, say, the Riemann conjecture. If '"mathematics" is not a sharply circumscribed concept' (MS 127, 185), but characterised by two key ideas, it is not surprising to find that in different cases they apply to different degrees: that beside paradigm cases of rigorously proven and applicable mathematics, there are instances of mathematics that are rigorously proven, but have little or no function as grammatical norms, and that we can also envisage less rigorously established mathematical propositions as long as they are usefully applicable.

'Constructivism' is a label that *can* be applied to Wittgenstein's philosophy of mathematics, as is clear from his famous slogan that 'the mathematician is an inventor, not a discoverer' (*RFM* I §168: 99), but he was obviously not a constructivist of an intuitionist type. He had no time for Brouwer's psychologism, the idea of mathematics as 'an essentially languageless activity of the mind' (Brouwer 1952, 141–2; cf. Chapter 7; Hacker 1986, 120–8; Marion 2003), nor would he agree with his revisionism, the demand that mathematics be reconstructed according to intuitionist principles. Wittgenstein's constructivism was not a foundationalist programme, but an attempt to describe what mathematicians actually do. They construct concepts, develop their implications, and in doing so sometimes modify and extend those concepts; not, to be sure, freely or arbitrarily, but restricted both by existing concepts and by concerns for the applicability of the resulting theorems (see Chapters 8–10).

Bibliography

Aigner, M. & Ziegler, G. (2001): *Proofs from the Book*, Berlin: Springer.

Ambrose, Alice (1959): 'Proof and the Theorem Proved', in: *Mind* 68; 435–45.

Anderson, Alan Ross (1958): 'Mathematics and the "Language Game"', in: P. Benacerraf & H. Putnam (eds.) (1964): *Philosophy of Mathematics. Selected Readings*, Englewood Cliffs, NJ: Prentice-Hall; 481–90.

Ayer, A.J. (1936): *Language, Truth, and Logic*, Harmondsworth: Penguin, 1971.

Baker, G.P. & Hacker, P.M.S. (1980): *Wittgenstein: Understanding and Meaning. An Analytical Commentary on the Philosophical Investigations*, vol. 1, Oxford: Blackwell.

Baker, G.P. & Hacker, P.M.S. (2009): 'Grammar and Necessity', in: G.P. Baker & P.M.S. Hacker (eds.), *Wittgenstein: Rules, Grammar and Necessity. An Analytical Commentary on the Philosophical Investigations*, vol. 2, 2nd ed., extensively rev. by P.M.S. Hacker, Oxford: Wiley Blackwell; 241–370.

Bangu, S. (2016): 'Later Wittgenstein on the Logicist Definition of Number', in: S. Costreie (ed.), *Early Analytic Philosophy. New Perspectives on the Tradition*. Volume in the Western Ontario Series in Philosophy of Science. Series editor W. Demopoulos, Wien & New York: Springer.

Bangu, S. (2017): 'Later Wittgenstein and the Genealogy of Mathematical Necessity', in: K.M. Cahill & T. Raleigh (eds.), *Wittgenstein and Naturalism*, London: Routledge; 151–73.

Bays, Timothy (2004): 'On Floyd and Putnam on Wittgenstein on Gödel', in: *Journal of Philosophy* CI (4); 197–210.

Becker, Oskar (1964): *Grundlagen der Mathematik in geschichtlicher Entwicklung*, Frankfurt/Main: Suhrkamp, 1975.

Benacerraf, P. (1965): 'What Numbers Could Not Be', in: *Philosophical Review* 74; 47–73.

Benacerraf, P. (1973): 'Mathematical Truth', in: *Journal of Philosophy* 70; 661–79; Reprinted in: Marcus, R. & McEvoy, M. (eds.) (2016): *An Historical Introduction to the Philosophy of Mathematics: A Reader*, London: Bloomsbury; 487–500.

Benacerraf, P. & Putnam, H. (eds.) (1964): *Philosophy of Mathematics. Selected Readings*, Englewood Cliffs, NJ: Prentice-Hall.

Bennett, Jonathan (1961): 'On Being Forced to a Conclusion', in: *Proceedings of the Aristotelian Society* 35; 15–34.

Bernays, Paul (1959): 'Comments on Ludwig Wittgenstein's *Remarks on the Foundations of Mathematics*', in: P. Benacerraf & H. Putnam (eds.) (1964):

Philosophy of Mathematics. Selected Readings, Englewood Cliffs, NJ: Prentice-Hall; 510–28.

Berto, F. (2009a): 'The Gödel Paradox and Wittgenstein's Reasons', in: *Philosophia Mathematica* III (17); 208–19.

Berto, F. (2009b): *There's Something About Gödel. The Complete Guide to the Incompleteness Theorem*, Oxford: Wiley Blackwell.

Bourbaki, N. (1939): *Eléments de mathématique: Théorie des ensembles*, Paris: Éditions Hermann, 1977.

Bouveresse, Jacques (1988): *Le Pays des Possibles: Wittgenstein, les mathématiques et le monde réel*, Paris: Les Éditions de Minuit.

Boyer, C.B. & Merzbach, U.C. (1989): *A History of Mathematics*, 2nd ed., New York: John Wiley.

Brouwer, L.E.J. (1952): 'Historical Background, Principles and Methods of Intuitionism', in: *South African Journal of Science* 49; 139–46.

Büttner, Kai (2016a): 'Surveyability and Mathematical Certainty', in: *Axiomathes* 26 (2) June 2016.

Büttner, Kai (2016b): 'Equinumerosity and One-One Correlatability', in: *Grazer Philosophische Studien* 93; 152–77.

Cantor, Georg (1891): 'Ueber eine elementare Frage der Mannigfaltigkeitslehre', in: *Jahresbericht der Deutschen Mathematiker-Vereinigung* 1; 75–8.

Chihara, C.S. (1977): 'Wittgenstein's Analysis of the Paradoxes in his *Lectures on the Foundations of Mathematics*', in: *Philosophical Review* 86; Reprinted in: Shanker 1986; 325–37.

Connes, Alain (2000): 'La réalité mathématique archaïque', in: *La Recherche* 332; 109.

Da Silva, Jairo Jose (1993): 'Wittgenstein on Irrational Numbers', in: K. Puhl (ed.), *Wittgenstein's Philosophy of Mathematics*, Vienna: Hölder-Pichler-Tempsky; 93–99.

Davis, Philip J. & Hersh, Reuben (1982): *The Mathematical Experience*, Brighton: The Harvester Press.

Dawson, Ryan (2016): 'Wittgenstein on Set Theory and the Enormously Big', in: *Philosophical Investigations* 39 (4); 313–34.

de Bruin, B. (2008): 'Wittgenstein on Circularity in the Frege-Russell Definition of Cardinal Number', in: *Philosophia Mathematica* III (6); 354–73.

Dehaene, Stanislas (2011): *The Number Sense: How the Mind Creates Mathematics*, Oxford: OUP.

du Sautoy, Marcus (2011): 'Exploring the Mathematical Library of Babel', in: J. Polkinghorne (ed.), *Meaning in Mathematics*, Oxford: OUP; 17–25.

Dummett, Michael (1959): 'Wittgenstein's Philosophy of Mathematics', in: P. Benacerraf & H. Putnam (eds.) (1964): *Philosophy of Mathematics. Selected Readings*, Englewood Cliffs, NJ: Prentice-Hall; 491–509 [also in: G. Pitcher (ed.) (1966): *Wittgenstein. The Philosophical Investigations*, London: Palgrave Macmillan; 420–47].

Feigl, Herbert (1981): 'The Wiener Kreis in America', in: R.S. Cohen (ed.), *Inquiries and Provocations: Selected Writings 1929–1974*, Dordrecht: D. Reidel; 57–94.

Floyd, Juliet & Mühlhölzer, Felix (2020): *Wittgenstein's Annotations to Hardy's Course of Pure Mathematics: An Investigation of Wittgenstein's Non-Extensionalist Understanding of the Real Numbers*, Cham: Springer.

Floyd, Juliet & Putnam, Hilary (2000): 'A Note on Wittgenstein's "Notorious Paragraph" About the Gödel Theorem', in: *The Journal of Philosophy* XCVII (11); 624–32.

Floyd, Juliet & Putnam, Hilary (2006): 'Bays, Steiner and Wittgenstein's "Notorious Paragraph" about the Gödel Theorem', in: *The Journal of Philosophy* CIII (2); 101–10.

Frascolla, P. (1994): *Wittgenstein's Philosophy of Mathematics*, London: Routledge.

Frascolla, P. (2001): 'Philosophy of Mathematics', in: H.J. Glock (ed.), *Wittgenstein. A Critical Reader*, Oxford: Wiley Blackwell; 268–88.

Frege, Gottlob (1884): *Foundations of Arithmetic*, trans. J.L. Austin, Oxford: Wiley Blackwell, 1950.

Frege, Gottlob (1889): 'Ueber die Zahlen des Herrn H. Schubert, Jena: Pohle', in: G. Patsig (ed.), *Logische Untersuchungen*, Göttingen: Vandenhoeck & Ruprecht, 1993; 133–62.

Frege, Gottlob (1893): *Basic Laws of Arithmetic*, trans. P.A. Ebert & M. Rossberg, Oxford: OUP, 2013.

Friederich, Simon (2011): 'Motivating Wittgenstein's Perspective on Mathematical Sentences as Norms', in: *Philosophia Mathematica* III (19); 1–19.

Galileo (1632): *Dialogues Concerning Two New Sciences*, Evanston, IL: Northwestern University, 1939. Reprinted by Dover 1954.

Gandon, Sebastien (2012): 'Wittgenstein et le logicisme de Russell: remarques critiques sur "A Mathematical Proof Must be Surveyable" de F. Mühlhölzer', in: *Philosophiques* 39 (1); 163–87.

Giaquinto, Marcus (2002): *The Search for Certainty: A Philosophical Account of Foundations of Mathematics*, Oxford: Clarendon Press.

Glock, Hans-Johann (1996): *A Wittgenstein Dictionary*, Oxford: Wiley Blackwell.

Glock, Hans-Johann & Büttner, Kai (2018): 'Mathematik und Begriffsbildung', in: J. Bromand (ed.), *Wittgenstein und die Philosophie der Mathematik*, Paderborn: Mentis; 175–93.

Gödel, Kurt (1931): 'Über formal unentscheidbare Sätze der Principia Mathematica und verwandter Systeme I, transl. On Formally Undecidable Propositions of *Principia Mathematica* and Related Systems I', in: S.G. Shanker (ed.), *Gödel's Theorem in Focus*, London: Routledge, 1989; 17–47.

Gödel, Kurt (1947): 'What Is Cantor's Continuum Problem?' in: P. Benacerraf & H. Putnam (eds.) (1964): *Philosophy of Mathematics. Selected Readings*, Englewood Cliffs, NJ: Prentice-Hall; 258–73.

Gödel, Kurt (2003): *Collected Works*, vol. 5, Oxford: OUP.

Goldstein, R. (2005): *Incompleteness: The Proof and Paradox of Kurt Gödel*, New York: Norton.

Goodstein, R.L. (1972): 'Wittgenstein's Philosophy of Mathematics', in: A. Ambrose & M. Lazerowitz (eds.), *Ludwig Wittgenstein: Philosophy and Language*, Bristol: Thoemmes Press, 1996; 271–86.

Gowers, Timothy (2006): 'Does Mathematics Need a Philosophy?' in: R. Hersh (ed.), *18 Unconventional Essays on the Nature of Mathematics*, New York: Springer; 182–201.

Hacker, P.M.S. (1986): *Insight and Illusion: Themes in the Philosophy of Wittgenstein*, rev. ed., Oxford: Clarendon Press.

Hacker, P.M.S. (1996): *Wittgenstein: Mind and Will: An Analytical Commentary on the Philosophical Investigations*, vol. 4, Oxford: Wiley Blackwell.

Hacking, Ian (2014): *Why Is There Philosophy of Mathematics at All?* Cambridge: CUP.

Han, Daesuk (2010): 'Wittgenstein and the Real Numbers', in: *History and Philosophy of Logic* 31 (3); 219–45.

Hart, H.L. (1961): *The Concept of Law*, Oxford: OUP.

Higgins, P.M. (2011): *Numbers*, Oxford: OUP.

Hilbert, David (1922): 'Neubegründung der Mathematik, Abhandlungen aus dem Seminar der Hamburgischen Universität, 1', English translation in P. Mancosu (ed.) (1998), *From Brouwer to Hilbert: The Debate on the Foundations of Mathematics in the 1920s*, Oxford: OUP; 198–214.

Hilbert, David (1925): 'On the Infinite', in: P. Benacerraf & H. Putnam (eds.) (1964): *Philosophy of Mathematics. Selected Readings*, Englewood Cliffs, NJ: Prentice-Hall; 134–51.

Hilbert, David (1928): 'Die Grundlagen der Mathematik', English translation in van Heijenoort (ed.), *From Frege to Gödel: A Sourcebook in Mathematical Logic, 1897–1931*, Cambridge, MA: Harvard UP, 1967; 464–79.

Hoffmann, D.W. (2017): *Die Gödel'schen Unvollständigkeitssätze*, 2nd ed., Berlin: Springer.

Intisar-ul-Haque (1978): 'Wittgenstein on Number', in: *International Philosophical Quarterly* 18; Reprinted in: Shanker 1986; 45–59.

Kant, Immanuel (1787), *Critique of Pure Reason*, trans. N. Kemp Smith, London: Palgrave Macmillan, 1929.

Kienzler, Wolfgang (1997): *Wittgensteins Wende zu seiner Spätphilosophie*, Frankfurt/Main: Suhrkamp.

Kienzler, Wolfgang (2008): 'Wittgensteins Anmerkungen zu Gödel. Eine Lektüre der *Bemerkungen über die Grundlagen der Mathematik*, Teil I, Anhang III', in: M. Kroß (ed.), *"Ein Netz von Normen": Wittgenstein und die Mathematik*, Berlin: Parerga; 149–98.

Kienzler, Wolfgang & Greve, Sebastian (2016): 'Wittgenstein on Gödelian "Incompleteness", Proofs and Mathematical Practice: Reading Remarks on the Foundations of Mathematics, Part I, Appendix III, Carefully', in: S. Greve & J. Macha (eds.), *Wittgenstein and the Creativity of Language*, Basingstoke: Palgrave Macmillan; 76–116.

Klenk, V.H. (1976): *Wittgenstein's Philosophy of Mathematics*, The Hague: Martinus Nijhoff.

Kline, Morris (1953): *Mathematics in Western Culture*, Harmondsworth: Penguin, 1972.

Kline, Morris (1980): *Mathematics: The Loss of Certainty*, New York: OUP.

Körner, Stephan (1960): *The Philosophy of Mathematics: An Introductory Essay*, London: Hutchinson.

Kreisel, G. (1958): 'Wittgenstein's Remarks on the Foundations of Mathematics', in: *British Journal for the Philosophy of Science* 9 (34); 135–58.

Kripke, Saul (1982): *Wittgenstein on Rules and Private Language*, Cambridge, MA: Harvard UP.

Lampert, Timm (2008): 'Wittgenstein on the Infinity of Primes', in: *History and Philosophy of Logic* 29; 63–81.

Lampert, Timm (2018): 'Wittgenstein and Gödel: An Attempt to Make "Wittgenstein's Objection" Reasonable', in: *Philosophia Mathematica* 26 (3); 324–45.

Leibniz, G.W. (1875–1890): *Die philosophischen Schriften*, ed. C.I. Gerhardt, Berlin: Weidmann. Reprinted by G. Olms.

Mancosu, Paolo (1998): 'Hilbert and Bernays on Metamathematics', in: P. Mancosu (ed.) (2010), *The Adventure of Reason: Interplay between Philosophy of Mathematics and Mathematical Logic, 1900–1940*, Oxford: OUP; 125–58.

Mancosu, Paolo (2004): 'Review of Gödel's *Collected Works*, Vols. IV and V', in: P. Mancosu (ed.) (2010), *The Adventure of Reason: Interplay between Philosophy of Mathematics and Mathematical Logic, 1900–1940*, Oxford: OUP; 240–54.

Mancosu, Paolo (2009): 'Measuring the Size of Infinite Collections of Natural Numbers: Was Cantor's Theory of Infinite Number Inevitable?' in: *The Review of Symbolic Logic* 2 (4), December; 612–46.

Mancosu, Paolo & Marion, Mathieu (2003): 'Wittgenstein's Constructivization of Euler's Proof of the Infinity of Primes', in: P. Mancosu (ed.) (2010), *The Adventure of Reason: Interplay between Philosophy of Mathematics and Mathematical Logic, 1900–1940*, Oxford: OUP; 217–31.

Marconi, Diego (2000): 'Verificationism in the *Tractatus*?' in: P. Frascolla (ed.), *Tractatus Logico-Philosophicus. Sources, Themes, Perspectives*, *Annali dell'Università della Basilicata*, vol. 11; 75–87.

Marconi, Diego (2002): 'Verificationism and the Transition', in: R. Haller & K. Puhl (eds.), *Wittgenstein and the Future of Philosophy. A Reassessment after 50 Years*, Vienna: Hölder-Pichler-Tempsky; 241–50.

Marion, Mathieu (1998): *Wittgenstein, Finitism, and the Foundations of Mathematics*, Oxford: Clarendon Press.

Marion, Mathieu (2003): 'Wittgenstein and Brouwer', in: *Synthese* 137; 103–27.

Marion, Mathieu (2008): 'Brouwer on "Hypotheses" and the Middle Wittgenstein', in: M.V. Atten, P. Boldini, M. Bourdeay & G. Heinzmann (eds.), *One Hundred Years of Intuitionism (1907–2007): The Cerisy Conference (Publications des Archives Henri Poincaré Publications of the Henri Poincaré Archives)*, 2008th ed., Basel: Birkhäuser Verlag; 96–114.

Marion, Mathieu (2011): 'Wittgenstein on Surveyability of Proofs', in: O. Kuusela & M. McGinn (eds.), *The Oxford Handbook of Wittgenstein*, Oxford: OUP; 138–61.

Marion, M. & Okada, M. (2013): 'Wittgenstein on Contradiction and Consistency: An Overview', in: *O Que No Faz Pensar* 33; 52–79.

Marion, M. & Okada, M. (2014): 'Wittgenstein on Equinumerosity and Surveyability', in: Kai Büttner et al. (eds.), *Themes from Wittgenstein and Quine*, Amsterdam: Rodopi; 61–78.

Marion, M. & Okada, M. (2018): 'Wittgenstein, Goodstein and the Origin of the Uniqueness Rule for Primitive Recursive Arithmetic', in: D. Stern (ed.), *Wittgenstein in the 1930s. Between the Tractatus and the Investigations*, Cambridge: CUP; 253–71.

McGuinness, Brian (1988): *Wittgenstein: A Life: Young Ludwig 1889–1921*, vol. 1, London: Duckworth.

Meschkowski, Herbert (1984): *Problemgeschichte der Mathematik I*, Mannheim: Bibliographisches Institut.

Monk, Ray (1990): *Wittgenstein: The Duty of Genius*, London: Jonathan Cape.

Mühlhölzer, Felix (2001): 'Wittgenstein and the Regular Heptagon', in: *Grazer Philosophische Studien* 62; 215–47.

Mühlhölzer, Felix (2006): '"A Mathematical Proof Must Be Surveyable". What Wittgenstein Meant by This and What It Implies', in: M. Kober (ed.), *Deepening Our Understanding of Wittgenstein*, Amsterdam: Rodopi.

Mühlhölzer, Felix (2010): *Braucht die Mathematik eine Grundlegung?* Frankfurt/Main: Klostermann.

Mühlhölzer, Felix (2012): 'Wittgenstein and Metamathematics', in: P. Stekeler-Weithofer (ed.), *Wittgenstein: Zu Philosophie und Wissenschaft*, Hamburg: Felix Meiner Verlag; 103–28.

Peacocke, C. (1981): 'Reply: Rule-Following: The Nature of Wittgenstein's Arguments', in: S. Holtzman & C. Leich (eds.), *Wittgenstein: To Follow a Rule*, London: Routledge; 72–95.

Poincaré, Henri (1905): *Science and Hypothesis*, trans. W.J. Greenstreet, London: Scott.

Potter, Michael (2011): 'Wittgenstein on Mathematics', in: O. Kuusela & M. McGinn (eds.), *The Oxford Handbook of Wittgenstein*, Oxford: OUP; 122–37.

Priest, Graham (2004): 'Wittgenstein's Remarks on Gödel's Theorem', in: M. Kölbel & B. Weiss (eds.), *Wittgenstein's Lasting Significance*, London: Routledge; 206–25.

Quine, Willard Van Orman (1936): 'Truth by Convention', in: P. Benacerraf & H. Putnam (eds.) (1964): *Philosophy of Mathematics. Selected Readings*, Englewood Cliffs, NJ: Prentice-Hall; 322–45.

Ramharter, Esther (ed.) (2008): *Prosa oder Beweis? Wittgensteins 'berüchtigte' Bemerkungen zu Gödel. Texte und Dokumente*, Berlin: Parerga.

Rodych, Victor (1999a): 'Wittgenstein's Inversion of Gödel's Theorem', in: *Erkenntnis* 51; 173–206.

Rodych, Victor (1999b): 'Wittgenstein on Irrationals and Algorithmic Decidability', in: *Synthese* 118; 279–304.

Rodych, Victor (2000): 'Wittgenstein's Critique of Set Theory', in: *The Southern Journal of Philosophy* XXXVIII; 281–319.

Rodych, Victor (2018): 'Wittgenstein's Philosophy of Mathematics', in: *Stanford Encyclopedia of Philosophy*, first published, 2007: https://plato.stanford.edu/entries/wittgenstein-mathematics/.

Rundle, Bede (1979): *Grammar in Philosophy*, Oxford: Clarendon Press.

Russell, Bertrand (1918): 'The Philosophy of Logical Atomism', in: *His: Logic and Knowledge. Essays: 1901–1950*, London: George Allen & Unwin, 1956; 175–282.

Russell, Bertrand (1919): *Introduction to Mathematical Philosophy*, London: George Allen & Unwin.

Russell, Bertrand (1956): *Portraits from Memory*, London: George Allen & Unwin.

Russell, Bertrand & Whitehead, A.N. (1927), *Principia Mathematica*, 2nd ed. (1st ed., 1910–13), Cambridge: CUP.

Ryle, Gilbert (1950): 'Heterologicality', in: *His Collected Papers*, vol. 2, London: Hutchinson, 1971.

Sainsbury, R.M. (1995): *Paradoxes*, 2nd ed., Cambridge: CUP.

Schroeder, Severin (1998): *Das Privatsprachen-Argument: Wittgenstein über Empfindung und Ausdruck*, Paderborn: Schöningh.

Schroeder, Severin (2001): 'Private Language and Private Experience', in: H.J. Glock (ed.), *Wittgenstein: A Critical Reader*, Oxford: Wiley Blackwell; 174–98.
Schroeder, Severin (2006): *Wittgenstein: The Way Out of the Fly-Bottle*, Cambridge: Polity.
Schroeder, Severin (2009a): 'Analytic Truths and Grammatical Propositions', in: H.J. Glock & J. Hyman (eds.), *Wittgenstein and Analytic Philosophy: Essays for P.M.S. Hacker*, Oxford: OUP; 83–108.
Schroeder, Severin (2009b): *Wittgenstein Lesen: Ein Kommentar zu ausgewählten Passagen der Philosophischen Untersuchungen*, Stuttgart-Bad Cannstatt: Fromman-Holzboog.
Schroeder, Severin (2012): 'Conjecture, Proof, and Sense, in Wittgenstein's Philosophy of Mathematics', in: C. Jäger & W. Löffler (eds.), *Epistemology: Contexts, Values, Disagreement. Proceedings of the 34th International Ludwig Wittgenstein Symposium in Kirchberg, 2011*, Frankfurt/Main: Ontos; 461–75.
Schroeder, Severin (2013): 'Wittgenstein on Rules in Language and Mathematics', in: N. Venturinha (ed.), *The Textual Genesis of Wittgenstein's Philosophical Investigations*, London: Routledge; 155–67.
Schroeder, Severin (2014): 'Mathematical Propositions as Rules of Grammar', in: *Grazer Philosophische Studien* 89; 21–36.
Schroeder, Severin (2015): 'Mathematics and Forms of Life', in: *Nordic Wittgenstein Review*, October; 111–30.
Schroeder, Severin (2016): 'Intuition, Decision, Compulsion', in: J. Padilla Gálvez (ed.), *Action, Decision-Making and Forms of Life*, Berlin: de Gruyter; 25–44.
Schroeder, Severin (2017a): 'On Some Standard Objections to Mathematical Conventionalism', in: *Belgrade Philosophical Annual* 30; 83–98.
Schroeder, Severin (2017b): 'Grammar and Grammatical Statements', in: H.J. Glock & J. Hyman (eds.), *A Companion to Wittgenstein*, Chichester: Wiley Blackwell; 252–68.
Shanker, S.G. (1986): 'Introduction: The Portals of Discovery' in: S.G. Shanker (ed.), *Ludwig Wittgenstein: Critical Assessments, Vol. 3: From the Tractatus to Remarks on the Foundations of Mathematics*, London: Croom Helm.
Shanker, S.G. (1987): *Wittgenstein and the Turning-Point in the Philosophy of Mathematics*, London: Routledge.
Shanker, S.G. (1988): 'Wittgenstein's Remarks on the Significance of Gödel's Theorem', in: S.G. Shanker (ed.), *Gödel's Theorem in Focus*, London: Routledge; 155–256.
Singh, Simon (1997): *Fermat's Last Theorem*, London: Fourth Estate.
Steiner, Mark (1975): *Mathematical Knowledge*, Ithaca, NY: Cornell UP.
Steiner, Mark (2000): 'Mathematical Intuition and Physical Intuition in Wittgenstein's Later Philosophy', in: *Synthese* 125; 333–40.
Steiner, Mark (2001): 'Wittgenstein as His Own Worst Enemy: The Case of Gödel's Theorem', in: *Philosophia Mathematica* 9 (3); 257–79.
Steiner, Mark (2009): 'Empirical Regularities in Wittgenstein's Philosophy of Mathematics', in: *Philosophia Mathematica* III (17); 1–34.
Steiner, Mark (2013): 'The "Silent Revolution" of Wittgenstein's Philosophy of Mathematics', presentation given at Göttingen University.
Stroud, Barry (1965): 'Wittgenstein and Logical Necessity', in: S. Shanker (ed.), *Ludwig Wittgenstein: Critical Assessments*, vol. 3, London: Routledge, 1996; 289–301.

Tarski, Alfred (1936): 'The Establishment of Scientific Semantics', in: *His: Logic, Semantics, Metamathematics: Papers from 1923 to 1938*, Oxford: OUP; 401–8.

Tarski, Alfred (1937): 'The Concept of Truth in Formalized Languages', in: *His: Logic, Semantics, Metamathematics: Papers from 1923 to 1938*, Oxford: Clarendon Press, 1956; 152–278.

von Savigny, Eike (1996): *Der Mensch als Mitmensch: Wittgensteins "Philosophische Untersuchungen"*, Munich: Deuscher Taschenbuchverlag.

Waismann, Friedrich (1936): *Einführung in das mathematische Denken*, München: dtv, 1970.

Waismann, Friedrich (1982): *Lectures on the Philosophy of Mathematics*, ed. W. Grassl, Amsterdam: Rodopi.

Wang, Hao (1961): 'Process and Existence in Mathematics', in: Y. Bar-Hillel et al. (eds.), *Essays on the Foundation of Mathematics dedicated to A.A. Fraenkel on His Seventieth Anniversary*, Jerusalem: Magnes; 328–51.

Weyl, Hermann (1921): 'Über die neue Grundlagenkrise der Mathematik', in: *Mathematische Zeitschrift* 10; 39–79.

Wright, Crispin (1980: *Wittgenstein on the Foundations of Mathematics*, Cambridge, MA: Harvard UP.

Wright, Georg Henrik von (1954): 'A Biographical Sketch', in: N. Malcolm (ed.), *Ludwig Wittgenstein: A Memoir*, Oxford: OUP, 1984; 1–20.

Wrigley, Michael (1980): 'Wittgenstein on Inconsistency', in: S. Shanker (ed.), *Ludwig Wittgenstein: Critical Assessments*, vol. 3, London: Routledge, 1986.

Wrigley, Michael (1989): 'The Origins of Wittgenstein's Verificationism', in: *Synthese* 78; 265–90.

Index

acid 167–8
agreement 80, 85, 88–92, 99–100, 107–8, 124, 134
algebra 51–4, 59, 151, 166, 191
Ambrose, Alice 150
analysis (conceptual) 3, 36, 41–2, 135–6, 172, 181n19, 191, 195, 196
analytic 32–3, 70, 93, 95–7, 102, 104–5, 134, 135, 168, 187, 210
anthropological approach x, 183, 196, 224
application of mathematics x, xi, 58, 60–4, 67–78, 116–21, 124–5, 126, 130, 134, 136–9, 140, 142, 144, 161n9, 173, 179, 181–8, 190, 191–2, 194, 198–202, 212n10, 214–15, 225
archive 106–7, 162
associative law of addition 51–3, 132n6, 150–1
augury 71–2
Augustine 4n1, 24, 220, 222
Ayer, A.J. 93, 102

Babylonians 128, 137, 162
Baker, Gordon 93, 114
Bangu, Sorin 131n5, 133n8
Barbier, Didier 202n4
belief 49–50, 66–7, 123, 173, 184, 223
Beltrami, Eugenio 77
Benacerraf, Paul 23n4, 221–2
Bennett, Jonathan 103
Bernays, Paul 217
Big Typescript 4, 47–50, 52–4, 141, 174, 177
Bipolarity principle 19
Bourbaki, Nicolas 202
Brouwer, L.E.J. 38n3, 82, 225
de Bruin, Boudewijn 20
Büttner, Kai 17–18, 19–21

calculation *see* proof
calculus (differential) 5, 7, 30, 39, 138, 189, 201
calculus model of language 181–2
calculus view x, 57–8, 129, 137–8, 141–2, 161n9, 170, 171, 183–4, 223
Cantor, Georg xi, 50, 151–8, 189, 206n2, 220; Cantor's Diagonal Proof 151–8
cardinality 156–8
Carroll, Lewis 136
certainty 5, 6–7, 30, 31–2, 33–4, 84–5, 189, 193
Chihara, Charles S. 193–6, 198–201
class paradox (Russell's paradox) 12–13
community 71, 88–92, 99, 106, 108, 117, 123
compulsion, logical 32, 83, 116, 117, 164–5, 175
concept 9, 25, 30, 49, 59, 61, 64, 65, 69, 70, 72, 77, 95, 108, 118, 122, 133, 136n9, 139–40, 147, 149, 150, 157, 159–60, 162–7, 173, 177, 185–6, 187, 225; change of 144–69, 177n18, 185–6, 216; connexion of 167; family-resemblance 80n2, 182; proclamatory 168
conjecture 37, 44–53, 115, 173–9, 184, 196, 217, 224, 225
consistency (inconsistency) 3, 5, 6, 7, 19n2, 26, 66–8, 69, 70, 77, 138, 160, 175, 180, 189–202, 225

construction, free 140
constructive 21–2, 159, 180
constructivism 225
contradiction 7, 12–13, 26, 70, 108, 124, 150, 158, 191–202, 216–18
convention 21, 30, 93–125, 137, 138, 145, 150, 162, 188
conventionalism 93–125, 126, 136, 138–9, 170; full-blooded/radical 105–9, 125
coronation 123
counting (transitive) 11, 20–1, 22–3, 25, 28–9, 63–5, 68–70, 72, 75–6, 107, 110, 116–17, 119–20, 125, 126, 128, 130–5, 136n10, 139–40, 144–5, 147, 157, 167, 187–8, 192, 200; *see also* number
creativity 140

decision 82–4, 106–9, 163–4, 168
Descartes, René 150
description 37, 46, 50, 53, 59, 62, 93, 110–11, 177–8, 181, 187, 190–1, 220–1, 224–5
Dolby, David 46, 157
Dualism, Cartesian 220–1
Dummett, Michael 83n5, 95, 101–3, 105–9
du Sautoy, Marcus 140

Egyptians 128, 137
Einstein, Albert 77
empiricism 59, 93, 95, 117, 119–20, 126–40, 142, 170
$_{N}$English (*vs.* $_{N+}$English) 69, 71, 74, 132, 187–8
equation 10, 14n6, 16, 25–8, 30–3, 35, 36–9, 41–5, 60–4, 66, 119, 123, 135, 136, 167–8, 171, 172
equinumerosity 8, 10–11, 13, 15–21, 23, 30, 143–6
essence 145, 148, 181
Euclid 76, 150, 158–66, 186, 217
Euclidian geometry 39, 75, 76, 77
Euler, Leonhard 76, 138
Euler's proof of infinity of primes 46
expansion (of irrational number) 151–5, 222–3
experiment 29, 30, 110, 119, 126–7, 131, 133, 142, 145, 169–70, 184, 186, 190; picture of an 169–70
extensionalism 50–1, 53, 151–4

Fermat's proposition/Last Theorem 43, 45, 47, 49, 166, 176, 184
Floyd, Juliet 118n10
formalism ix, x, 3, 4, 189–90, 225
form of life 71, 124–5, 200
foundation(s) viii, xi, 3–8, 12, 13, 21n3, 27–34, 82, 87, 108n6, 133n7, 189, 200, 202, 203, 225
Franz Josef I. 162
Frascolla, Pasquale 91–2
Frege, Gottlob ix, 5, 7–8, 9–12, 15–25, 26, 27, 35, 81, 83, 101–2, 132n6, 135–6, 189, 193, 201, 210
Friederich, Simon 191n1

Galileo 158
Gandon, Sébastien 31n6
Gauß, Carl Friedrich 159, 216n17
generality (generalisation) 38n3, 43, 50, 51–2, 53, 59, 67, 80–1, 97, 102, 106–9, 119–20, 127–8, 130, 132, 133n8, 137, 144, 146, 151, 153, 162, 170, 178, 181, 182n21, 191
God 71, 224
Gödel 5, 158n8, 197, 203–19, 222, 224; incompleteness theorems 203–19, 222, 224
Goldbach's conjecture 49–50, 176, 178
Gowers, Timothy 23, 50n12, 221, 222–4, 225
grammar 4, 39, 42, 48, 59–77, 78, 118n10, 187, 203, 220, 225; autonomy of 130–1
grammar view x, 57–77, 118n10, 130, 136, 137, 141, 142, 161n9, 171, 172, 184, 187, 223
grammatical proposition (rule) x, 39–41, 53, 58, 59–77, 78, 93, 95, 96, 98, 126–8, 135, 136, 141–2, 145, 161, 171, 176, 177, 182, 184, 192, 215, 223
Greek mathematics 138

Hacker, P.M.S. 93, 114
Hacking, Ian 213n11
heptagon, construction of 159–60, 177–8, 215–16
Hilbert, David x–xi, 7, 189–93, 211, 225
Hilbert's Hotel 158

Hume, David 146n1
Husserl, Edmund 145

identity 10, 14n6, 16–17, 25, 79, 135, 136, 144, 168, 203
inconsistency *see* consistency; contradiction
inference 31–3, 35, 60, 95, 97, 100, 101–8, 109–15, 123, 131, 190
infinity x, 5–6, 13, 39, 45, 50–1, 53, 76, 79, 97, 103–4, 151, 154, 157–8, 178, 179, 180, 189, 190, 220; axiom of 13, 27
inner object model 221n1
instrumentalism 125, 126
interpretation 75, 78, 80, 81, 85–6, 190, 192, 207–19
Intisar-ul-Haque 23n4
intuition 81–4, 99, 145, 190, 222
intuitionism ix, 3, 4, 38n3, 82, 84, 189, 225

jigsaw puzzle 146, 149, 161–2, 175
Jourdain, Philip 13

Kant, Immanuel 64, 93, 135
Kepler's Second Law 138
Kepler's sphere-packing conjecture 50
Kienzler, Wolfgang 206n1, 217, 219n18
Kline, Morris 5
know-how 87–8
Kraus, Karl 200
Kreisel, Georg 61–2
Kripke, Saul 91, 105, 108–9
Kronecker, Leopold 50

Lagrange, Joseph Louis de 138
language game 142–3, 181–3, 186–7, 221
law (legal system) 142, 162, 168, 171, 176, 192; *see also* rule
Leibniz, Gottfried Wilhelm 132n6, 158
Liar Paradox 196, 201, 202, 209–10, 211–12, 217–18
logic 9–34, 35, 36–7, 53, 75, 95–105, 109–15, 133, 135, 189, 194–6, 200, 203, 213; *see also* inference
logical empiricism/positivism 93–5, 101
logicism ix, 3, 4–5, 6–7, 9–34, 136n9, 189, 201, 210
Lueger, Karl 162

Mancosu, Paolo 158
marbles 169–70
Marion, Mathieu 20–1, 31, 50, 53n15
meaning 4n1, 7, 22, 23–5, 47, 61, 64–5, 66, 69, 72, 75–6, 88–9, 93–4, 98, 99, 100, 102–3, 105, 108–9, 110, 118, 131, 133n7, 141, 145, 150, 174, 176–80, 181, 182, 183, 184, 187, 208, 218–19, 220–1
measurement 39, 66–71, 75, 109–14, 122, 127–8, 159, 162, 177–8, 179, 187
mental representation 108–9
meta-mathematics 190–1
Mill, John Stuart 59
Mühlhölzer, Felix 118n10, 165

necessity 61, 69, 70, 76, 93–4, 96–7, 101, 103, 104, 105–6, 108, 118n10, 126, 131, 132, 133n8, 138, 146, 175
Newton, Isaac 138
non-denumerability 156–7
non-Euclidian geometry 77, 199
normativity 53, 61, 70, 74, 76, 89, 92, 94, 97, 99–100, 106, 136n10, 142, 147, 161n9, 171, 172, 176, 177n18, 178, 182n21, 191n1, 197, 215
norm of representation/expression 58, 66–70, 77, 120, 124, 126–7, 142, 170, 171, 173, 176, 223–4; *see also* grammatical proposition
number: definition of 7–12, 13, 14, 15–25, 26–7, 29, 93, 132, 152, 154–5, 187; irrational 151–8, 189, 223; Mersenne prime 76, 186; perfect 76, 186; prime 46–7, 49, 76, 135, 160–6, 175, 179, 180, 217; real 151–8; transfinite 189, 190; *see also* counting (transitive)

Okada, Mitsuhiro 20–1
one-to-one correlation 11, 15–21, 76, 143–6, 148, 153, 157–8
overlap-addition 115–16, 139, 147, 186

part-whole principle 158
Peano's axioms 23n4, 133n7
pentacle, pentagon 143–6, 148; construction of regular pentagon 158–60, 177
Pepys, Samuel 90

Philosophical Investigations viii, x, 3, 79–80, 89
Philosophical Remarks 36–47, 174
Platonism x, 22–5, 32, 50, 53, 59, 67, 81, 83, 94, 108n6, 117–18, 145, 191, 203, 213, 220–5
Poincaré, Jules Henri 52, 136n9
practice (activity) 21, 62, 69, 74–6, 78, 84, 87–92, 98, 114, 117, 122–3, 124, 131, 133, 139, 141, 142, 183, 186, 197, 223–5
prediction 62–4, 68, 71–2, 73, 74–5, 133–4, 184, 187–8
Prince, Derek 198–9
Principle of Explosion 193–6
prison 197
private 89–90, 220
problem, mathematical 3, 5, 37, 39, 42–50, 53, 177–8; *see also* conjecture
proof 5, 7, 12, 27–31, 35, 36n2, 37–54, 57–8, 61–2, 73, 75, 76, 77, 115–16, 117n9, 120–1, 129, 130, 131, 133n8, 134, 135, 137, 138, 141–88, 203–19, 223–4; H-P 143–6, 148; inductive (recursive) 51–4, 150–1, 156, 190; provability 209, 210–13, 216, 219, 222–3, 225
proof theory *see* formalism
prose, mathematical 45–8, 174, 191

quantifier 9–10, 25, 28, 30–1
quantity 37, 70, 72–4, 109, 121–3, 134, 141, 192, 195, 199–200
Quine, W.V.O. 95–100, 101n3, 103, 105, 107

referentialism 6–7, 23–5, 221–2
relation 11, 15–18, 36, 46n10, 75, 102–3, 110, 119, 139, 168, 172, 205, 211–12
religion 6, 71, 141–2, 179, 183, 224–5
research, mathematical 8, 42–50, 53, 77, 172–8
Riemann, Georg Bernhard 77; conjecture 225
Rodych, Victor 14n6, 153, 155, 206–8
Rogers, C.A. 50
rule: empirical propositions hardened into 126–40, 170, 223–4; of grammar (*see* grammatical proposition); substitution 25, 64, 102, 167, 171; transformation 35–6, 39, 182
rule-following paradox 43, 53, 78–92, 105–9, 115–16, 117–18, 153, 163–5
Russell, Bertrand ix, 4, 5, 6–7, 10n3, 12–13, 15–23, 26–34, 35, 81, 83, 184, 189, 200–1, 203, 210
Russian Peasant Multiplication (Ethiopian multiplication) 129–30, 131

sample 24, 70, 79, 107, 110, 112
Schlick, Moritz 196
search, mathematical 43n8, 44, 46n11, 172, 178, 193, 196, 200, 216–17
self-reference 205–10, 211–12, 214, 217
self-verifying/self-evident 35–6, 47
sense 35–6, 37–41, 42–4, 47–8, 59, 62, 66, 67, 71, 74, 93, 142, 159, 168–9, 171–2, 173–80, 181, 183, 184, 187, 218–19; determinacy of 41, 172, 182; *see also* meaning
set theory x–xi, 34, 50, 53, 189, 200, 203, 210, 220, 221n2
Shanker, Stuart 53n15, 140
Skolem, Thoralf 51–3, 150–1, 206n2
Steiner, Mark 31n6, 130–4, 136n10, 142
stipulation 39, 49, 52, 61, 94, 131–2, 144–51, 153, 157, 161–3, 192; *see also* concept, change of
Stroud, Barry 122n13
surveyable 29–31, 144
synthetic a priori 52, 93, 134–6

Tarski, Alfred 211
tautology 27–8, 35, 36–7, 41, 133, 135, 204
theology 71, 72n10
theory 189–91, 199
theory of types 13, 26–7, 201
Tractatus Logico-Philosophicus viii, ix, x, 13–14, 19, 35–6, 38n3, 50, 51, 53, 60, 181
trisecting an angle 166
truth-functor 33
Turing, Allan 193–202

understanding 5, 7–8, 33, 38, 41, 57, 80, 86–91, 104–5, 116, 164–5, 172, 174, 175, 180, 184, 216
use x, 4n1, 8, 10, 23–5, 33, 35, 49, 58–61, 67–8, 70, 72, 75, 80n3, 85,

88, 91–4, 96, 98–100, 105, 107–8, 118, 122–3, 141, 155, 161, 167, 172–3, 179, 182–3, 185, 186–7, 191–2, 196, 220–1

vagueness 25, 40, 156, 165, 178, 182
verification 21–3, 27, 70, 171, 175, 179, 181, 186–7, 190
verificationism 37–8, 53, 57–8, 172, 179, 181, 183–4, 187
Voltaire 138

Waismann, Friedrich 196
Wang, Hao 30–1
Wantzel, Pierre 166
Weierstrass, Karl 7
Weyl, Hermann 38n3, 43
Whitehead, Alfred North 7, 201, 203
Wiles, Andrew 45, 166, 176
wood sellers 121–3, 200
Wright, Crispin 95, 103–5, 108–9, 112–14, 122, 123
Wrigley, Michael 101